The Unknown Universe

ALSO BY THE AUTHOR

NON-FICTION

The Sun Kings

Deep Space

Galaxy

Voyager

Is There Life on Mars?

FICTION

The Sky's Dark Labyrinth

The Sensorium of God

The Day Without Yesterday

The Unknown Universe

A New Exploration of Time, Space, and Cosmology

Stuart Clark, Ph.D.

PEGASUS BOOKS
NEW YORK LONDON

THE UNKNOWN UNIVERSE

Pegasus Books Ltd.
80 Broad Street, 5th Floor
New York, NY 10004

First Pegasus Books hardcover edition July 2016

ISBN: 978-1-68177-153-3

10 9 8 7 6 5 4 3 2 1

Printed in the United States of America
Distributed by W. W. Norton & Company, Inc.

Contents

The Unknown Universe

Introduction
THE DAY WE SAW
THE UNIVERSE

It was the day that cosmologists had been waiting for. The day when we were promised the ultimate picture of the early Universe. It was 21 March 2013, just twenty-four hours after the spring equinox, and the symbolism could not have been richer. As the bleakness of winter was giving way to the hope of spring, we were entering a new season of cosmology, one when the answers to the origin of the Universe were laid bare before our eyes.

They were contained in a single image that would be revealed by the European Space Agency (ESA) at a press conference in its Paris headquarters. It had been taken by an ESA spacecraft stationed 1.5 million kilometres away from Earth. Called Planck, after the great German physicist Max Planck, the probe had spent two and a half years painstakingly building up the picture, pixel by pixel.

The image showed what the sky would look like if our eyes could see microwaves instead of light. At first glance, it does not look special, just a mottled oval of blues and golds, yet it is arguably the most important image of the Universe ever taken.

Fundamentally, microwaves and visible light are the same thing. They're both waves that carry energy through space. The only difference is their wavelength. Microwaves are about ten thousand times longer than visible light. Not that this makes them huge. The microwaves captured on the image had all been somewhere between 0.3 and 11.1 millimetres in length, and had been travelling through space for almost 14 billion years. They were part of the very first rays of 'light' to be generated in the Universe and had begun their journey across space more than 9 billion years before our planet had even formed. Although images had been taken of this radiation before, none were as detailed as the Planck image; none held the promise of such insight into our cosmic origins.

For reasons that I will discuss in Chapter 8, the microwaves prove that the cosmos had a beginning, or at the very least was once drastically different from today. Yet the joke of it is that these signals were dismissed as crap when they were first detected by a pair of American radio engineers in 1964. Literally, crap.

Arno Penzias and Robert Wilson were tinkering with an old radio receiver that had fallen into disuse. They were sharing it with a pair of homing pigeons that had taken up residence and had covered it in droppings (the pigeons not the engineers).

When the telescope picked up an all-pervading static hiss, Penzias and Wilson blamed it on electrical interference from the pigeon poo. They removed the birds by driving them across the state and releasing them. Then they cleaned the telescope.

But what do pigeons do? Yes, they fly home to roost. So the problem returned and Penzias and Wilson had to find a different solution. This time they came up with a permanent solution, involving a man and a shotgun. They cleaned the telescope once more and started observing again.

The hiss remained.

It was not the droppings but the Nobel Prize-winning discovery of the cosmic microwave background radiation, the oldest light in the Universe. As such, it became a vital component of cosmology, the science of understanding the Universe. It was emitted just 380,000 years after the Universe was supposed to have come into being during the mysterious event that astronomers have dubbed the Big Bang.

As an archaeologist digs down through older and older layers of the Earth to see the pattern of evolution, so astronomers look further and further away. The further they set their gaze, the longer it has taken light to cross that distance, and so the older the celestial objects they see. As we shall see in Chapter 4, light travels quickly but not with infinite speed. In a single year, unhindered it can cross 9.5 thousand billion kilometres and this is the distance that astronomers call a light year. If something were 1 light year away, its light would have taken a year to reach us, and so we would be seeing it as it appeared one year ago, when it emitted the light. There would be no way to know what it looks like now. It's like receiving a letter that has been lost in the post and wondering if its message is still current.

The upside is that it allows astronomers to study the changing nature of the cosmos. For example, think about our cosmic neighbourhood. It extends to a few hundred light years away. So stars at the outer edge of this region appear to us as they did when Europe's Age of Enlightenment was at its height. The nearest large star-forming cloud of gas, the Orion Nebula, is some 1300 light years away. It appears as it did in the seventh century AD, when the people of the Arabian peninsula were first united under the prophet Mohammed and began the spread of Islam.

At 158,000 light years away, a modest nearby collection of stars called the Large Magellanic Cloud looks as it did when the earliest humans were still confined to the African continent. The light from the Andromeda Galaxy, the nearest large collection of stars, began its journey across space 2.3 million years ago, when the *Homo* lineage that became humans was just diverging from the great apes. Another prominent galaxy called Centaurus A sits 13 million light years away. Its appearance is roughly coincident with the evolution of the great apes on Earth. Studying the succession of these objects allows us to track the way things have changed in our Universe.

The cosmological microwave background radiation itself, with its 13.7 billion year vintage, represents the earliest view of the Universe we could ever get at these wavelengths. There were no planets or stars at that time, just a gigantic cloud of atoms filling the whole Universe. The mottling on the Planck image reveals the subtle variations in density across this cloud. As the cosmic clock ticked on, gravity pulled each denser region more tightly together, eventually giving birth to the first stars. In a very real sense, the Planck image can be thought of as nothing less than the blueprint for our cosmos.

And Planck's instruments are working at the limits of physics, not technology. In other words, it is practically impossible to build better instruments. In terms of seeing the Universe's microwave blueprint, this image is essentially the best that humankind is ever going to get. So how do we use it?

The Universe we live in today is a hierarchy of shining structures. Stars are gravitationally bound to each other in rotating collections known as galaxies. Galaxies are gravitationally bound to each other in collections known as clusters, and the clusters are strung through space in filaments that make up the cosmic web. All of this magnificence grew from the minute density variations present in the microwave background.

These variations are therefore the essential starting point for computer programs, called models, that mimic the evolution of the Universe. Crudely, the trick is to take the microwave pattern and see if our understanding of physics can transform it into the cosmic web of today's Universe.

The models themselves are mathematical recipes that take the laws of physics as their foundation and then add the 'ingredients' of the Universe. For cosmology, gravity is the essential law of physics. There are three other forces of nature (we will encounter electromagnetism in Chapter 4 and the two nuclear forces in Chapter 7) but these play only minor roles in shaping the Universe overall.

The model's ingredients are given by six parameters. The first two are measured from the mottling in the microwave background. Parameter one is their amplitude: in other words, how great the variations in the gas density are across the Universe. The second is to do with the volume of space in which these variations occur. Some are small volume fluctuations, others are much larger. This parameter measures the difference in amplitude between the smallest and the largest volumes.

Then we come to the contents of the Universe. A central theme of this book will be the path trodden by cosmologists in their attempts to define the average density of matter and energy in the Universe. It has proved to be anything but easy. To make their models work with any semblance of success, they have been forced to assume that the ordinary atoms making up the stars, planets and life consist of no more than 4 per cent of the total contents of the Universe. The other 96 per cent of everything is in forms of matter and energy that are unknown to us. Worse than this, the calculations show that they are almost beyond our ability to detect directly. They call the unknown stuff dark matter and dark energy, and they infer its existence by measuring the movement of galaxies.

The majority of galaxies appear to be rotating too quickly, or moving away from us in space at an ever-accelerating rate. Hence, the cosmologists believe that they need dark matter to spin the galaxies faster, and dark energy to push them away from us more quickly. These three ingredients – atoms, dark matter and dark energy – can be summarized in just two parameters because they are all dependent on one another. If you know the proportion of any two, the other can be simply deduced.

The fifth parameter of the standard model of cosmology is related to when the first stars formed. This point in cosmic history lies beyond the reach of even our best telescopes as yet. It was a catastrophic event in which almost every hydrogen atom in the Universe was ripped apart because the newly formed stars released huge amounts of destructive ultraviolet light. It occurred after the release of the cosmic microwave background and determines how easy it is for the microwaves to travel uninterrupted through space.

The sixth and final parameter is the Universe's expansion rate. This is known as Hubble's constant, after the American astronomer Edwin Hubble, who published definitive evidence of the expansion of the Universe in 1929 (see Chapter 8).

In a perfect scenario, cosmologists would measure each of these parameters using completely independent means, plug them into the model, and out would come an answer that perfectly matched the distribution of galaxies in today's universe. In reality, it is not that easy. Some of the parameters can be measured; others have to be estimated.

Then there are the assumptions, such as the existence of dark matter and dark energy, and the little mathematical fudges that have to be put into the model to turn it into a calculation that can be solved. If one of these is wrong, then the model itself is wrong and what we thought we knew about the Universe would evaporate before our eyes.

Having said that, confidence in the standard model took a huge leap forward thanks to the work of a NASA spacecraft called WMAP, the Wilkinson Microwave Anisotropy Probe. It was a forerunner to Planck and launched in June 2001. The word 'anisotropy' is the technical term for the density variations across the early Universe, and for nine years WMAP repeatedly observed these. It hugely improved the accuracy of the standard model's first two parameters, and as a result improved the accuracy of the model by a factor of more than 68,000.

On the face of it, there seemed little doubt that the standard model must be substantially correct, and cosmologists began to trumpet their victory. The WMAP website lists ten achievements that follow from the use of WMAP data and the standard model. From the age of the Universe to the percentage of ordinary atoms, cosmology was said to have entered an era of 'precision'. What was not mentioned on the website's list of achievements were the data that the standard model struggled to explain.

WMAP had seen a hint that the mottling in one part of the sky was deeper than the standard model allowed. It was dubbed 'the cold spot' because the anisotropies can be translated into temperatures but the detection was so slight that some thought it could have been a bit of instrumental noise.

So a key question was: had Planck seen it too?

There were also more general concerns about the ingredients of the standard model: namely dark matter and dark energy. After decades of theoretical work and experimentation, no one has been able to conclusively detect a single piece of dark matter. As we will discuss in Chapter 7, the hints we have from the various detectors around the world are both confusing and contradictory.

The dark energy is even more mysterious. There is no natural candidate that springs from any physics we currently understand. Some of our current hypotheses, such as particle physics supersymmetry (see Chapter 7), were designed specifically to exclude such an energy. So, perhaps dark matter and dark energy are not real. Perhaps they are phantoms conjured into being by a deeper misunderstanding of the Universe. If so, the standard model will have to be replaced.

Yet none of these concerns were voiced by NASA astrophysicist and Nobel laureate John Mather. On the eve of the ESA press conference, he was quoted by the BBC as saying: 'I'm hoping there's something surprising there for them. If they just say, "Well, other people were right" – that's not exciting; the last decimal places are never very interesting. What we want is some new phenomenon.'*

Mather had won the 2006 Nobel Prize in Physics for his work on the microwave background radiation using a NASA spacecraft called COBE, the COsmic Background Explorer. A year later, *Time* magazine listed him as one of the 100 most

* http://www.bbc.co.uk/news/science-environment-21828202

influential people in the world. Now, he was in charge of the biggest space mission in the world today, the NASA-led James Webb Space Telescope, with its eye-watering $8 billion price tag. However, you looked at it, his opinion carried real weight.

It was a public reflection of an undercurrent I had encountered several times. A number of cosmologists had given me 'off-the-record' comments that Planck was a waste of money because WMAP had effectively allowed cosmologists to extract all the really useful information from the microwave background. The implication was clear: more precision would simply confirm what WMAP had already found.

The irony of Mather's statement was in his dismissal of the 'last decimal places'. He had shared the Nobel Prize with cosmologist George Smoot for their discovery of the cosmic blueprint, as revealed by the temperature anisotropies in the microwave background. Those anisotropies had been found in the last decimal places it was possible to extract from the data they had been using.

The temperature of the gas in the Universe back then was around 3000 °C, whereas the blueprint is encoded in variations that are on average just 20 millionths of a degree from place to place. Yet, from this imperceptible temperature variation had sprung the galaxies, which each now contained between hundreds of thousands and hundreds of billions of individual stars.

Far from being irrelevant, the last decimal places to which you can measure often contain the most interest, because there you see the hints of what you don't understand – all those tricky details that remain to be explained. The last decimal places are the reason scientists always want bigger, better, more precise technology.

More and more detailed observations are the bedrock of true science. They tell us what the Universe is actually like, not what a theoretician calculates it should be like on average. And in twenty-four hours, the world would know.

Nerves were on edge when the ESA press conference began. Those who could not attend in person watched via a live stream on the Internet. Twitter was abuzz.

To signal just how important the event was to the agency, the director general of ESA, Jean-Jacques Dordain, spoke first. In sombre tones and broken English, he said that Planck had revealed an 'almost perfect' universe. But what did he mean by 'almost perfect'?

He left it to Professor George Efstathiou, of the University of Cambridge, UK, to explain. One of the foremost cosmologists, Efstathiou once held the same position in Oxford as Edmond Halley, the famous seventeenth-century astronomer.

At the beginning of the press conference, Efstathiou looked tense. His lips were pressed together into a thin line, his shoulders were hunched. When he started talking, the tension disappeared; he seemed at ease and fluent, speaking precisely, almost downbeat. He announced without fanfare that the screen now showed the most precise map of the microwave background that had ever been obtained. It was a gold mine of information, he said, even though 'it may look a little like a dirty rugby ball or a piece of modern art'.

No one laughed and he ploughed on, assuring the audience that there were cosmologists who would have 'hacked our computers or maybe even given up their children to get hold of a copy of this map'. Still no one laughed.

He said that the Planck map was incredibly exciting, but

instead of saying why, he then gave a lecture on basic cosmology. Almost half an hour into the press conference, nothing new had been said. When he presented the conclusions it was little more than small tweaks to what was already known. There was about 5 per cent ordinary matter instead of 4 per cent, the proportion of dark matter to dark energy was a little different, the Universe was 80 million years older than we thought, making it 13.8 billion years rather than 13.7 billion. The overall conclusion, he said, was that the standard model of cosmology is an extremely good match to the Planck data.

Watching from my office at home, I was poised at the keyboard to write up the results for *Across the Universe*,* my astronomy blog hosted on the *Guardian* newspaper's website, and I was starting to feel anxious. I received an email from a friend, a senior UK science editor, saying, 'If this is all they are going to say, this is a nightmare.'

Indeed, John Mather's worst fears were coming true before our eyes.

Then it all changed. Efstathiou said, 'But there are some issues, and that is why we have described the science results as an almost perfect Universe.'

He began to stumble on his words; he looked down while he was speaking. He reiterated how good the standard model was at fitting the data, and added that he could have simply stopped there and said 'cosmology is finished'. But rather hesitantly he pushed himself to say, 'But because we've got such good fit to the data [overall], we should examine more critically what doesn't seem to fit. We have to look at what hasn't fitted because that is where there may be evidence of new physics.'

* www.theguardian.com/science/across-the-universe/2013/mar/21/european-space-agency-astronomy

At last, the game was afoot. Here were the 'new phenomena' that Mather (and the rest of us!) wanted. We were about to step into the unknown.

Efstathiou explained that, on the largest scale of the Universe, the temperature fluctuations were smaller than expected and that such behaviour was impossible in the standard model of cosmology. Also, the average temperature fluctuations on one side of the sky were larger than on the other; again, that was forbidden by the standard model. Finally, as the accompanying press release* confirmed, but Efstathiou did not mention, the WMAP 'cold spot' had been seen, confirming its existence.

The quality of the detection removed any doubts about the reality of these anomalies. They were all real features of the primordial universe – and they were impossible to understand with standard thinking. There was no tweak that the Planck team had tried that could explain where these features were coming from. The message, according to Efstathiou, was that the Planck data showed 'cosmology is not finished'.

In February 2015, Chuck Bennett, professor of physics and astronomy at Johns Hopkins University, and colleagues conducted a thorough comparison of the cosmological model derived from WMAP with that from Planck.† Worryingly, they found that the two solutions are not consistent with each other – each described a different Universe. Clearly something is amiss somewhere. The two might not have been exactly correct, but they should have been consistent. The error is now under investigation: either one of the data

* www.esa.int/Our_Activities/Space_Science/Planck/Planck_reveals_an_almost_perfect_Universe

† http://arxiv.org/abs/1409.7718v2

sets has been calibrated incorrectly or the standard model is wrong.

But how can we make progress when the Planck image is just about the very best we can obtain of the microwave anisotropies, our primary source of information about the early Universe?

For all our achievements, do we yet live in an unknown universe waiting to be explored and understood?

Frankly, Douglas Adams could not have written it any better. It was the world's 42 moment for real. Most cosmologists thought that the answer to Life, the Universe and Everything (by which I mean the origin of the Universe) would become clear from the Planck data, yet right now no one really knew what to make of it.

The majority think that all these little snags are merely the final details to be clarified, a little bit of scientific 'i'-dotting and 't'-crossing, but a growing number think that they are signs that we are completely wrong about the Universe.

It is into those uncharted realms that this book will journey. The search for answers will take us into the most mysterious places in the Universe; it will take us into the hearts of black holes, the moment of the Big Bang, and to a confrontation with the very nature of reality itself.

And it all starts on England's Great North Road, between London and Cambridge, in the latter decades of the seventeenth century.

I

The Architect of the Universe

It was August 1684, twenty years after the Restoration of the English monarchy and the novelty of Charles II was wearing thin. The roads were in poor repair and the summer heat would not have helped; the ground must have been cracked, the air choked with dust. Yet this probably seemed the least of Edmond Halley's problems as he made his way from London to Cambridge.

Two years shy of his thirtieth birthday, Halley should have been in an enviable position. He was recognized as one of the foremost astronomers of his day, and able to indulge his passion for the stars by living off his father's money. The family business was making soap, and the Halleys had become unexpectedly wealthy in the years following the Great Plague of 1665–6, when London society began to find washing fashionable.

His father's wealth bought Edmond an education at St Paul's School in London, and then a place at Queen's College, Oxford. There, Edmond grew into a handsome man of intelligence, with an eye for the ladies and a dash of the reckless adventurer about him. He made his name by walking out of his degree, claiming that the old-fashioned syllabus was suffocating, and winning the patronage of the king to chart the southern stars from the tropical Atlantic island of St Helena. Such a map was of prime military importance in those days because the navy needed accurate star charts to find its way at sea.

Halley carried out his duty with aplomb, returning eighteen months later with the positions of 341 stars. He even named a constellation *Robur Carolinum* (Charles's oak),* after the tree in which his monarch was said to have hidden from Oliver Cromwell's troops following the Battle of Worcester in 1651.

His star chart placed Halley on the map as well. Soon, he was travelling across Europe, visiting the great astronomers to discuss results and observing techniques. In 1682, he observed the comet that would one day bear his name, thanks to a groundbreaking calculation of its orbit he would perform in 1704. But in 1684, it was not comet orbits that were driving him to distraction but the motion of the planets. Indeed, it was searching for a solution to this conundrum that set him on the road to Cambridge that summer, and set the world on course for a completely different view of the cosmos and a whole new way of gathering knowledge.

Until this point in history, the Universe was thought to have been a pretty simple place. There was the Sun, the Earth and the Moon; these were the most obvious of the celestial bodies. There were the stars, which appeared night after night, fixed in their constellations. Then there were the objects of delicious mystery: the five planets visible to the naked eye. Mercury, Venus, Mars, Jupiter and Saturn chased each other across the sky at their own individual paces, propelled by unknown powers. Indeed, it was this movement that led to the Greeks calling them *planetes* (wanderers).

* The constellation was not included in the official list of constellations drawn up by International Astronomical Union in 1922. *Robur Carolinum* was located near the modern constellations of Crux and Carina.

Little more was known about them apart from their different colours and intensities. Mars was a fiery red. Jupiter and Venus were glittering white points, but whereas Jupiter was frequently seen in the fully dark sky, Venus never strayed far from twilight. In this behaviour, it was joined by the dimmer point of Mercury, which positively shunned the darkness. Saturn, on the other hand, was no stranger to the deep night; its baleful, yellow countenance became synonymous with gloom, giving us the adjective 'saturnine'.

The question of how the planets moved remained unanswered for millennia. Some thought legions of marching angels were responsible, but that raised the larger question of why. If there was one implicit assumption in all of this, it was that the heavens were an orderly place where things happened for a reason. So what purpose did the restless orbs of the heavens serve?

Astrology, a system of belief based around divining the future, grew to fill this void. In this scheme, the planets were thought to rain influences down on Earth, affecting our mood and tweaking our personalities. Saturn, as mentioned, was responsible for gloomy thoughts of apathy. Jupiter, on the other hand, was associated with the spirit of the hunter. Everything – from passion to music or warfare – had its origins in the planets. Where the celestial wanderers were at the moment of our birth somehow imprinted itself upon us, making us susceptible to those personality traits. The higher the planet in the sky, the more we were affected. In many ways, astrology was a forerunner of psychology, in that it sought to explain our personalities and why we behave the way we do.

By comparing the current position of the planets with their positions at the moment of a client's birth, an astrologer could advise on an individual's reaction to future events and therefore suggest the best course of action to take: perhaps by waiting until an unhelpful planet had passed from sight, or a counterbalancing one had risen.

The German mathematician and astronomer Johannes Kepler (1571–1630) made much of his living as an astrologer. To make his divinations, he had to know as precisely as possible where the stars and the planets were. This was where astronomy came in, with its star charts and planetary ephemerides. Yet the true nature of the planets and the reason for their orbits remained entirely unknown because the astrologers simply didn't need that information. Most were simply not interested in such knowledge, but, if push came to shove, it was said that the heavenly bodies were the work of God and therefore beyond the comprehension of dim-brained humans.

Kepler was different. He wondered about the nature of the astrological interactions. He even speculated on whether houses could be built with special shielding to insulate the inhabitants from the tug of these (often unhelpful) planetary influences. Such thinking helped drive his research into the nature of the planets.

Working in the early decades of the seventeenth century, Kepler fought what he later called his 'war with Mars'. It took him more than a decade to achieve victory.

Like others, he was both deeply religious and convinced that the Universe was a place of order. But he went a step further by believing that the order was mathematical and that geometry could be used to uncover nothing less than God's design for the cosmos.

In pursuit of that design, he spent years trying to determine the size and shape of Mars's orbit by poring over the observational data amassed by the great Danish astronomer Tycho Brahe. His analysis was complicated by the fact that the Earth, following its own orbit, was a moving platform. So his first job was to remove Earth's motion from the calculations. Once he did that, what remained was the true movement of Mars: an elliptical orbit around the Sun. Emboldened by this triumph, he moved on to the other planets and found that they, too, follow their own ellipses.

The results of his toil remain spectacular even today. After more than a decade of effort and number crunching, he discovered that the movement of the planets could be described by just three mathematical laws. These laws of planetary motion are still one of the first things taught on a modern astronomy course.

The first of Kepler's laws is that planets move in elliptical orbits around the Sun. The second is a precise mathematical description of how a planet speeds up when close to the Sun and how it slows down when it is far away. The third law relates the average speed of a planet to the size of its orbit, showing that more distant planets move at a slower speed.

Beyond the science that this achievement would spark, the cultural value was immense. Through the act of measuring the positions of the stars, Tycho Brahe had captured nature and transformed it into numbers. Then Kepler had used mathematics to distil that information – hundreds of pages of raw data – into something meaningful: a precise description of planetary motion that could be written in just three simple lines of mathematics. To cap it all, Kepler had done this for the one realm that most thought was beyond human comprehension: the heavens.

It was a staggering call to intellectual arms. Observations

and mind power could unlock the secret of places no human could yet visit, only see.

But for all the success, there were still questions. Kepler could describe the movement of the planets but say nothing about why they moved. His intuition told him that some influence was coming from the Sun but the mathematical nature of it escaped him. For all his triumph, it had merely cleared the entrance to a pathway, rather than completed the journey. Decades later, Edmond Halley found himself heading down that same path – and he wasn't alone.

In London, a group of gentlemen dedicated to the investigation of natural knowledge had banded together under a royal charter from Charles II. They were known as the Royal Society of London for Improving Natural Knowledge. Today the organization is known merely as the Royal Society, but the motto remains the same: *nullius in verba*. Roughly translated, it means 'take no one's word for it' and encapsulates the rule that personal belief is not enough in science. Knowledge cannot be accepted until it is shown by measurement to be consistent with the natural world.

In January 1684, Halley had met with two other fellows of the Royal Society: astronomer and architect Christopher Wren and experimentalist and surveyor Robert Hooke. Both of these men had been driving the reconstruction of London following the Great Fire of 1666, but that night they were talking about what drove the planets to obey Kepler's laws.

By simply tinkering with mathematics, all three had hit upon the idea that Kepler's third law, which related the average speed of a planet to its average distance from the Sun, implied that a force was acting in a very specific way:

an inverse square. This means that if you double the distance of a planet from the Sun, the intensity of the force drops by a quarter; if you triple the distance, the force drops to a ninth of its original value. But the challenge was mathematically proving that such an inverse square law would drive the planets into elliptical orbits.

As the three philosophers huddled in the aromatic warmth of Jonathan's coffee shop, buried in the tangle of alleyways near the heart of the City of London, Hooke claimed to be able to do that but refused to show his workings until others had tried and failed. Wren offered him 40 shillings for the proof, but Hooke remained implacable.

Why he behaved like his is lost in the mists of time. Perhaps the quarrelsome experimentalist ensnared himself in a moment of boasting. Certainly, the proof is a far from simple calculation and Hooke was not an overtly skilled mathematician. Halley was much more numerically competent, yet the task escaped him too, and this led him to Cambridge in August 1684. He went to ask Isaac Newton for help.

At this time, Newton was a shadowy figure. He was far from the public figure that he became in later life. He shunned London society and the burgeoning discipline of natural philosophy for an altogether shadier pursuit: alchemy. In his darkened rooms at Trinity College, Newton sat for days tending his furnace. Bathed in noxious fumes he stirred the chemicals, hoping to discover the secrets of transformation.

The goal of the true alchemist was not the pursuit of wealth but the pursuit of the Philosopher's Stone. This was the hypothetical elixir said to transform one substance into another, so changing a base metal into gold was just one example of its power. It was also thought to be the key difference between living things and inanimate matter. This was what Newton craved: the power of life over death.

It would be sufficiently dramatic to say that Halley's visit changed the course of Newton's life, but the truth is that Halley's visit that summer changed the course of history. The world we live in today is built upon the work that Newton performed in the wake of their conversation. Yet, as Halley walked into Newton's room at Trinity College, neither of them could have known the realms of wonder that were to open before their eyes.

According to the account given by Newton to his friend, the French mathematician Abraham de Moivre, Halley simply asked him a question: what would be the path of a planet under the influence of an inverse square law? To this, Newton responded with the answer: an ellipse, as in Kepler's first law of planetary motion. Halley was struck with 'joy and amazement' and asked for the proof. After searching his papers, Newton claimed to have lost the workings.

We can only imagine Halley's disappointment. Here was a second man saying that he had performed the work, yet was incapable of producing the proof. Unlike Hooke, Newton pledged to recalculate and send the paper down to London, so Halley left, empty-handed. Months passed and with each new turn of the calendar Newton's promise must have looked increasingly unreliable.

What Halley did not know about was the mental effort being exerted in Cambridge. Alchemy had been abandoned and Newton was in thrall to natural philosophy as never before.

After a false start in which he wrecked the calculation by mislabelling a set of axes on a hastily drawn diagram, Newton began to make progress. He completed the derivation of

Kepler's first law but saw that there was more to do – much more. By November, however, he was ready.

The document he sent to Halley contained derivations of all three of Kepler's laws, and a generalization of the first law to show that the precise shape of an orbit was determined by the average velocity of the celestial object. Ellipses were just one possibility; other mathematical shapes were possible too.

Stunned by what he read, Halley immediately returned to Cambridge, seeking Newton's permission to circulate the contents of the document. Newton refused because he claimed that the work was not finished. In calculating why the planets moved as they did, he had glimpsed an entirely new science of why things moved – not just planets, but everything.

Newton had calculated the planetary motions precisely, without needing to account for friction or any other form of resistance to movement. Yet, if he included such things, other trajectories became obvious – trajectories that looked a lot like the motion of falling objects on Earth. His curiosity inflamed, the Cambridge don fell upon the subject like a bear upon its prey. He fired off letter after letter to other philosophers asking for observations and measurements that he could use to test his new ideas. He wanted the times of the tides at Deptford, the observations of the planets Jupiter and Saturn when they drew close in the sky, and the paths of comets that had graced the night.

In one such letter to Astronomer Royal John Flamsteed, Newton declared that his work would remain secret until it was finished to his satisfaction. 'I would gladly know the bottom of it before I publish my papers,' he wrote.

It took him nearly three years. During that time, he virtually placed himself under house arrest and did little but work. He was sharing rooms with his amanuensis Humphrey Newton at the time. Although they were not related, Humphrey came to know Isaac more closely than any family member. He witnessed the great philosopher's compulsive behaviour and described how Isaac was so consumed with work that he ate only sparingly, often forgetting to eat at all, and how he would absent-mindedly go about the simplest daily tasks because his head was so full of his studies. At other times, he would shriek like Archimedes and rush to his desk, where he would stand and scribble furiously, neglecting to pull up a chair.

By 1687, Newton's masterpiece, the *Philosophiæ Naturalis Principia Mathematica* (*Mathematical Principles of Natural Philosophy*) was complete. The philosopher and the work emerged into a changed England.

Charles II had died and his brother, the Catholic James II, had been crowned by a reluctant Protestant government. James embodied the threat of Vatican rule, the mere thought of which made England uneasy. There was political confusion and the spectre of civil war. Uncertainty was rife; chaos beckoned.

Yet to those who could understand – and, just as importantly, believe in – Newton's labyrinthine mathematics, the *Principia* was a beacon of certainty. The work showed how motion was always the product of a force. If something moved, the underlying force could be calculated. Sometimes the force was obvious; it could be an ox pulling a cart, a person throwing a ball into the air, or a mass of men pushing on a wall to topple it. At other times, the force was invisible, such as when an object drops to the floor, a pebble rolls downhill, or a planet passes along its orbit. Newton showed that the invisible force was the signature of gravity. As astonishing as it may seem, rain falls to Earth for the same reason that a comet slips through the night sky: because gravity is acting on it.

There was a price to pay for this grand insight. In Newton's concept, gravity was a force that acted across emptiness without the need for physical contact. He called this 'action at a distance' and had taken his inspiration from alchemy, which had bred in him a belief that spirits pervaded the cosmos and could be channelled by an alchemist to help the experiments. Alchemy was thought to be the boundary between the material and the spiritual world, with the alchemist's state of mind capable of influencing the outcome of the experiments.

In effect, Newton saw gravity as one of these spirits but, having found its correct mathematical description, realized that it was impervious to mental influence. The only factors that affected its strength were the masses of the two objects involved and the square of the distance between them. No matter how hard Newton concentrated, his mind would not alter the force of gravity. It was an utterly materialistic force of nature.

The problem was that action at a distance struck many as completely unscientific. Most natural philosophers had adopted the notion of nature as a mechanism and that meant movement could only be achieved through contact.

To illustrate this, imagine standing at the end of a line of books and wanting to push over the furthest copy without walking to it. The only way to achieve the goal is to topple the nearest book so that the line tumbles like a set of dominoes until the force is transported to the far end. To explain gravity, natural philosophers had postulated the ether, a fluid-like substance, to sweep the planets through their orbits like a river carrying leaves.

Newton's equation, however, did not support the need for the ether. Indeed, when he tried to analyse planetary motion as a swirling vortex, with friction and resistance, the numbers did not add up. The only thing that worked was ephemeral action at a distance, but that meant no one knew what gravity was or how it worked. It reminded some people too much of astrology and mystical influences from the heavens, so they dismissed the concept on those grounds alone. In drawing such battle lines, they allowed the true power of mathematical science to be demonstrated once and for all, because Newton's formula for gravity allowed predictions to be made and tested with observations. If the predictions were shown to be true, then regardless of whether anyone liked the idea or not, it had to be accepted as the truth – and that was where Halley and his comet came in.

The young astronomer had been so entwined in the formulation and publication of Newton's *Principia* that without him the work would never have been produced. He had acted as editor and publisher and had even bankrolled the project because the Royal Society had exhausted its finances publishing *The History of Fishes*, a 'sure-fire winner' that had utterly failed to set the book-buying public alight. Indeed, the society was so straitened by the flop* that Halley, who was their secretary, was once paid in copies of the deathless fish tome to the value of the monies owing to him.

Comet orbits proved almost as implacable as *The History of Fishes*. Little enough is known about these celestial objects today but back in the seventeenth century they were entirely enigmatic. For centuries they were thought to be atmospheric phenomena that foretold coming disasters.

* No pun intended.

In the sixteenth century, German painter and astronomer Georg Busch published an extraordinary commentary that claimed comets 'are formed by the ascending from the earth of human sins and wickedness, formed into a kind of gas, and ignited by the anger of God. This poisonous stuff falls down again on people's heads, and causes all kinds of mischief, such as pestilence, sudden death, bad weather and Frenchmen.' Really.

One person who rounded on Busch for these absurd, unsupported views was Tycho Brahe. In the welter of celestial observations that he and his assistants had taken in the latter decades of the sixteenth century, he had collected observations of a bright comet that appeared in 1577. These observations showed that the comet was moving through space, not Earth's atmosphere. Hence, Brahe dismissed Busch's opinions as merely 'figments and pigments' and presented calculations that showed the comet to be at least four times further than the Moon.

Although that was almost certainly a severe underestimate, the implication by Newton's time was clear: comets were space-borne objects, therefore subject to the pull of the Sun's gravity and so should be in orbit around it.

To prove this, Newton analysed two comets. The first had appeared in the evening skies of 1680; the second in the morning sky of 1681. He reasoned that they were the same object travelling on a highly elongated orbit around the Sun and found that the best shape to fit the comet's path was a tight curve known as a parabola. But the calculation was so laborious that Newton was not keen to repeat the exercise for other comets.

In 1704, Halley took up the challenge. He deduced the

orbits for twenty-three comets that had appeared between the years 1337 and 1698. Three struck him immediately. The comets of 1531, 1607 and 1682 had all followed similar paths to one another. The orbits were not identical but they were close enough to raise his suspicion.

His moment of genius came when he remembered to apply Newton's principle of 'universal gravitation'. This was the second of Newton's great insights, after action at a distance, and was the one that Newton claimed to have received when he saw an apple fall from a tree. That momentary event – so he claimed – set him thinking about the extent of gravity's reach. He wondered how much force you would need to shoot something straight up without it ever falling back to Earth.

As he investigated this mathematically, he realized that gravity extends away from celestial objects and weakens in accordance with the inverse square law, but it never reaches zero. Indeed, the pull of gravity is essential to guide planets, moons and comets through their orbits.

Everything pulls on everything else and this is what Newton called universal gravitation. For comets sweeping past planets in their fall towards the Sun, Halley reasoned that these passing interactions would have a calculable effect on their orbit. So he looked up the positions of Jupiter and Saturn, the two largest planets, and calculated the perturbing effect of their gravity on the comets of 1531, 1607 and 1682. Suddenly, it all made sense: it was the same comet that had been seen three times. The orbital differences were entirely down to the deflecting effect of Jupiter and Saturn.

He calculated when the comet was likely to return, then modified his result to include the effects of the two giant planets, and published his prediction in Latin in the Royal

Society's journal, *Philosophical Transactions*. He published it again a year later, this time in English for good measure. The comet would return, he said, in 1758–9.

German farmer and amateur astronomer Johann Georg Palitzsch saw it on Christmas Day 1758. Halley was right; Newtonian gravitation was right. What all the soothsayers and astrologers had claimed to do, Halley had now actually done: he had predicted a future event. If you had to choose a moment when science was truly born, this could very well be it.

The achievement was a convincing argument that the natural world was a place of order that could be laid bare by measurement and mathematical investigation. What Kepler had glimpsed, so Newton had uncovered, and Halley had proved: mathematical analysis of nature could determine the cause of events, and reveal the order of the Universe.

By using a mathematical hypothesis to make a firm prediction about an as-yet-unknown event, science established its defining characteristic: falsifiability. This is the ability for a scientific hypothesis to be tested, and shown to be wrong if it fails to match reality. Newton's prescription for a scientific investigation, often referred to as the scientific method, was that a mathematical hypothesis should be formed from a set of measurements taken of the natural world. This hypothesis should then be used to make a unique prediction that could be tested. If the prediction was shown to be true, either by experimentation or further observation, then the hypothesis was renamed a theory. However, this theory must never just be believed: it had to prove its worth continually by continuing to be tested.

With his work on comets, this is exactly what Halley had done, but the astronomer himself never lived to see his vindication. On 14 January 1742, after a twenty-two-year stint as the second Astronomer Royal, the eighty-five-year-old Halley asked for a glass of red wine. He drank it, placed the glass on the table next to his chair, and expired peacefully.

The predicted return of Halley's comet fitted perfectly with the goals of the Enlightenment. Sometimes called the Age of Reason, this was a cultural movement that was sweeping Europe at the time and sought to reform society by the use of reason rather than tradition.

There were three routes that could be taken in the pursuit of knowledge. Newton's work embodies the theoretical approach, in which existing observations are explained by the revelation of an underlying principle that can be expressed in mathematical form. Such a theory tells us *why* something happens.

Others in the Royal Society, such as Newton's arch-rival Robert Hooke, worked to an experimental agenda. Here, natural phenomena are isolated or even reproduced using apparatus in a laboratory to study them at close quarters. They can then be described mathematically and the resultant equations are known as empirical laws and describe *how* something happens.

Both of these approaches constitute science as we would understand it today. During the Enlightenment there was also the third way: the rationalist approach. This has largely fallen from favour today but allowed philosophers to develop possible explanations for the world using reason alone.

One of the greatest Enlightenment philosophers was Immanuel Kant. In this new age of reason, Kant and others felt the time was right to investigate mentally how the planets had formed.

Perhaps the most striking aspect of the planets is their orderly arrangement. Each crosses the night sky at a different rate but follows essentially the same path. This is known as the ecliptic and cuts through the twelve zodiacal constellations – Aries, Taurus, Gemini and so on.

It is a stunning piece of order, especially when compared to the comets, which arrive from myriad directions. This was enough to convince Newton that the planets could not have fallen into the same plane by accident. Instead, he saw the arrangement of the Solar System as the hand of God at work.

Kant and the other philosophers pursued a different belief. To them, order could ensue from blind physics and they attempted to envisage the sequence of natural events that could lead to the planets all forming in the same plane. It did not mean that they were working towards an atheistic agenda. Most thought that God was the ultimate reason for the Universe's existence but that the day-to-day course of things was simply a matter of physics. This fitted with the idea of nature as a mechanism, because a clock or other mechanical device must first be created before it can be left to work on its own.

In this new age of reason, it was no longer enough to think that God had created the planets in the biblical Genesis. Having now understood the power of mathematics to unlock nature's secrets, the intellectuals wanted to investigate and understand the process by which everything had happened, including the formation of the planets.

The first scientific hypothesis about their origin had come decades before from the French philosopher René Descartes. In a book published posthumously in 1677, he reasoned that the planets were trapped in a vortex of ethereal particles centred on the Sun. The circulation of the particles swept the planets into orbits, like leaves on a river being guided by the current. He also thought it likely that the planets had formed from the condensation of these particles.

Although Newton had shown mathematically that fluid motion in a vortex could not reproduce Kepler's laws of planetary motion, the idea of planets condensing out of 'the ether' persisted in the mind of philosophers. It was an elegant, almost obvious picture, and during the eighteenth century, those smitten with the idea began the process of modification.

Swedish philosopher Emanuel Swedenborg (1688–1772) suggested that the planets were condensations of matter thrown out from the equator of the Sun. But it was Kant who struck gold in his book *Allgemeine Naturgeschichte und Theorie des Himmels* (*Universal Natural History and Theory of the Heavens*). Kant's scenario for the formation of the planets was called the nebular hypothesis, after the Latin word for cloud, *nebula*. He took his inspiration from some puzzling astronomical observations.

Throughout history, various astronomers had noted the appearance of 'smudges' in the night sky. Although the study of these 'nebulous stars' began in earnest with the invention of the telescope in 1610, progress at first was slow.

In 1656, Dutch astronomer Christiaan Huygens drew attention to the Orion Nebula, a patch of nebulosity that appeared to the naked eye as a pinkish star in Orion's sword, but through a telescope was magnified enough to reveal its true nebular status.

In 1715, Edmond Halley published a list of six he had discovered at the eyepiece, and thereafter the numbers steadily rose. More and more of them were found as the century progressed. Some appeared to be stars wrapped in misty veils; other looked truly cloud-like with no obvious condensation. They lit Kant's imagination and he envisaged them as gradually falling together, drawn in by gravity, until enough gas accumulated at the centre and a star lit up. The planets were separate, smaller condensations that took place around the central star.

Later in the century, the great French scientist Pierre-Simon Laplace (1749–1827) expanded on the idea. He concluded that the collapsing cloud must be rotating and that this would force it to spin into a disc from which the planets then formed. He speculated that rings of material would break off from this gradually contracting disc, and so the outer planets were older than the inner ones. He was wrong on the latter point, but he was spot on in the way the cloud pancakes into a disc.

It happens because of centrifugal force. This is not a fundamental force such as gravity but is created by a spinning object and experienced as a feeling of being pushed away from the centre of rotation. The faster something rotates, the greater the centrifugal force felt. It is the reason a car will leave the road if the driver takes a bend too fast.

For a gas cloud rotating in space, gravity is pulling it together to form a star and centrifugal force is resisting this, but mostly around the cloud's equator where the rotation is greatest.

Along the rotation axis at the poles, there is hardly any movement and so little or no centrifugal force. Hence, the cloud collapses faster along the axis than through the equator and this makes it pancake into a disc. Within this disc, smaller gravitational condensations become the planets.

The scenario was greeted with interest, but to have any scientific currency, it had to stand up mathematically. This was where the astronomers and their calculations came in. This is also where the problems started.

The chances are that we have all watched an ice skater spinning during a performance. The closer to her body she tucks her arms, the faster she spins. The Sun should do the same. As it tucked in more and more gas from the collapsing cloud, it should have spun faster. Yet today it rotates quite leisurely: just once every month or so. In comparison, the planets positively zip around their orbits, even though they should be moving quite sluggishly because they were formed in the outer parts of the cloud. This seemed the exact opposite of what was expected from the theory.

By the beginnings of the twentieth century, astronomers had turned away from the nebular hypothesis and filled the void with other ideas, some outlandish.

One idea was that the Sun passed through an interstellar cloud, snagging a veil of gas that became the planets. However, this could not really explain why the Sun was rotating so slowly. Another idea proposed that the Sun originally possessed a companion star that was struck by a third, passing star. The collisions created the debris that then condensed into the planets. A third idea ameliorated this catastrophe by saying that a passing star's gravity had ripped out the raw ingredients for the planets from the Sun.

In truth, nothing worked. The maths did not add up and the scenarios were Byzantine compared with the simplicity of the nebular hypothesis. But how could astronomers believe in that any more?

It took until the 1980s and a pioneering space telescope

to find a way forward. As impossible as the nebular hypothesis sounded, one observation in the middle of the decade convinced astronomers to look again.

It was only 6 p.m. but the great Pacific night was already fully formed. It was 25 January 1983 and the only sound rolling across the launch complex was the rhythm of the waves from the nearby shoreline. On the pad, a 35-metre-tall rocket pointed to the stars, waiting for the countdown to complete. At zero, the darkness was interrupted by a blinding flash of light and the giant clawed its way into the sky above California's Vandenberg Air Force Base. As it lifted itself to heaven, the heat from its exhaust temporarily banished the winter cold.

Cradled in the nose cone was IRAS, the Infrared Astronomical Satellite. A joint mission between the United States, the United Kingdom and the Netherlands, it was designed to show astronomers the Universe in a different kind of light: something invisible to humans.

Beyond ordinary light is the infrared region of the electromagnetic spectrum. We cannot see this radiation but instead perceive it as heat. IRAS had infrared eyes and could see the warmth from celestial objects like a security camera can see body heat.

The satellite surveyed the whole sky, looking for any celestial object that gave out significant quantities of infrared, and in some places it found much more than its masters had bargained for. Several stars, including one named Beta Pictoris, because it is the second-brightest star in the constellation Pictor (the Easel), gave out a truly excessive quantity of infrared. The amount was so much more than expected from the star that astronomers began to scratch their heads.

The solution became obvious in April of the following year, when Bradford Smith, from the University of Arizona, and Richard Terrile, from NASA's Jet Propulsion Laboratory, targeted Beta Pictoris from Chile's desert-bound Las Campanas Observatory. Using a specially constructed mask inside the telescope, the pair blotted out the light from the centre, rather like when you raise your hand to block out the dazzling glare from the Sun. Then they took an image.

It was disappointing; nothing much showed up at all. But when they loaded the image onto their computers back in California and began a routine analysis, things immediately took a dramatic turn.

There was definitely something around the star. Two needle-like projections cut through the star, diagonally opposite one another. It was so obvious that the astronomers thought they had introduced it by making an error in the computer processing. After days of trying different ways of processing the image, the feature remained and the astronomers began to believe it was real. They then looked at other images of the star that they had taken with the instrument aligned in different ways, yet always the feature remained and in the same orientation. It had to be real.

It was a disc of matter surrounding the star and appeared as a thin line because it just happened to be seen edge-on from our point of view. From the size of the disc on the image, Smith and Terrile estimated that it contained enough matter to build about 200 planets the size of the Earth.

Further calculations showed that the likely temperature of this dust was enough to produce the infrared excess that IRAS had detected. Yet the truly astonishing thing was that the disc was exactly as Swedenborg, Kant and Laplace had reasoned two centuries earlier. Human reasoning had made a prediction that had been shown to be true. Even though there were still problems with the angular momentum, the discovery galvanized astronomers into looking for solutions.

In the subsequent years, more and more dust discs were discovered around other young stars; so, too, was a gale-force wind of atoms that the stellar youngsters throw off in their final phases of formation. Known as the T Tauri wind, after the first star from which it was discovered, it was a first step in understanding how a star reduces its angular momentum. The wind travelling outwards is like the ice skater opening their arms again to naturally reduce the rate of rotation. There is very likely a deceleration caused by the magnetic field of the star as well, but at the time, Smith and Terrile were not thinking that far ahead. They had another, more pressing concern. They wanted to see the dust disc with their own eyes; images no longer seemed good enough.

So they built a special mask that would provide an image to their eye, rather than a camera. They did this over breakfast, with cardboard and thread glued together with grape jelly from the observatory's canteen and their own saliva. They took the mask to the eyepiece that night and fulfilled their ambition to be the first humans to see a circumstellar disc.

Spurred on by this cardinal discovery and many subsequent observations, astronomers have now built a working hypothesis for our Solar System's formation. It is that the planets and their attendant moons were born 4.5 billion years ago, out of a multitude of mountain-sized rocks called planetesimals. They formed a giant asteroid belt that stretched from close to the Sun to hundreds of times the distance of Earth.

In this crowded environment, collisions were frequent. With everything moving in more or less the same directions, the impacts tended to be relatively gentle. Instead of shattering, head-on events, the energy of the impact would have melted the planetesimals, allowing them to coalesce and resolidify as a single, larger object.

Being molten, the denser material would have sunk to the centre of the body, producing metallic cores, while the lighter rocks floated to the outer layers. Over the course of half a billion years, the swarm of planetesimals coalesced into today's family of planets and moons. By the latter stages of this process, the major bodies were assembled and the remaining planetesimals continued to collide, excavating craters. It is these craters that are seen on the surface of the Moon today.

It is an elegant scenario, backed up by observations and calculations that mean it has to be close to the truth. Reason, theory and observation all show this to be true. But not everything is solved, and the greatest puzzle of all is right here on our celestial doorstep: the Moon.

2 Selene’s Secrets

It was perhaps the wildest off-road drive of them all, and it hadn't started well. A test outing the day before had resulted in a broken steering system, requiring some hasty repairs before the main event: a seven-hour safari with no roads and only one-sixth the gravity of Earth.

When astronauts David Scott and James Irwin climbed into their 'jeep' and set off, they could not have known that, seven hours later, their day's work would be described as 'the greatest day of scientific exploration that we've ever seen in the space programme – possibly of all time'.

They were on the Moon as part of the Apollo 15 mission and the clock was ticking.

Funding was being slashed as the American government had realized just how much money it cost to go to the Moon. NASA was now in a rush. The Apollo 15 astronauts needed to perform experiments originally planned for the now-cancelled Apollo 20 mission. The whole programme was going to close after Apollo 17. As a result, the remaining astronauts were receiving crash courses in geology. By learning how to classify rocks, they would be able to scout for the most valuable specimens.

In addition, the first task at each working site was to set up cameras that could be controlled from Earth. That way, geologists back home could also help them look.

The rocks that had been brought back by the first astronauts had been interesting in their own right but were largely ordinary. Mostly they were jumbled fragments of rocks smashed during the impacts that made the lunar craters. What geologists really sought was something truly ancient that would date back to the Moon's formation. This would help them determine the Moon's raw ingredients and, by extension, those of the planets as well.

If the planets and moons had formed from the collision of smaller rocky objects, as was looking increasingly likely, then the Moon would once have been entirely molten. The lightest minerals would have floated to the top of this global magma ocean and crystallized into rocks known as anorthosites. This was what Scott and Irwin were looking for – so far, however, without any luck.

After several hours of exploring, they parked the lunar rover near the rim of a small crater, set up the camera and equipment, and began to scout around, collecting as they went. Scott noticed it first: a gleaming white rock that stood out from the surrounding grey ones. Irwin pronounced it 'a beaut'.

Scott picked it up and began to clean off the dust from its surface. As he did so, sunlight glinted from the crystalline surface, causing both astronauts to exclaim.

'Guess what we just found?' Scott teased mission control to Irwin's laugh of pleasure. 'Guess what we just found! I think we found what we came for.'

'Crystalline rock, huh?' said Irwin.

'Yes, sir. You better believe it,' said Scott before going on to say that he thought it was something 'close to anorthosite'.

When the sample was returned to Earth on 7 August

1971, in a splashdown just north of Hawaii, it was swiftly discovered that the rock was indeed anorthosite: almost a pure sample. The press began referring to it as the 'Genesis Rock', assuming that it was a piece of the very first lunar crust.* To mark the importance of the find, Irwin had a replica made, which he would carry around in his briefcase or coat pocket as a talking point for many years afterwards.

In laboratories behind the scenes, work immediately began on unlocking its secrets and one thing became obvious: Moon rocks were surprisingly similar in composition to Earth rocks. This was a shock to some researchers because it spoke of a common heritage, that somehow the Moon and the Earth had once been joined. But how?

For many decades, the most popular scenario for the formation of the Moon had come from George Darwin, son of the famous naturalist Charles. Drawn to astronomy rather than natural history, George became the Plumian Professor of Astronomy at the University of Cambridge, a post whose statutes had been written by Isaac Newton in 1707.

By the time George assumed the chair in 1883, it was a position of privilege. Also, by this time, science was turning its attention to the composition of the Earth. Way back in 1798, Henry Cavendish had measured the world's average density to be 5.4 times that of water, yet it took almost a century for the implication of this work to be acknowledged.

The fact of the matter was that the average density of the Earth was almost twice that of ordinary rocks. To produce

* It has been subsequently discovered through radioactive dating that the sample does not quite reach back to the 4.5-billion-year-old primordial crust but crystallized a few hundred million years later, probably in the aftermath of a giant impact that melted a fraction of the Moon's surface.

such a result meant that there must be some far denser material deep inside the Earth, and the only thing that could be was metal.

Meteorites had been discovered and analysed that were almost purely iron, and so it seemed reasonable to believe that this and other heavy metals would be found inside the Earth, perhaps forming a central core.

Darwin set about calculating how the early Earth would have responded to the formation of such a compact, heavy core. He treated the planet as if it were the (now proverbial) ice skater mentioned in Chapter 1, and calculated the way in which the rotation rate of our planet would speed up as the heavy metals sank to the centre of the planet.

His work showed that just as the ice skater speeds up when tucked into a more compact configuration, so too would the Earth. Maybe the acceleration was enough, he reasoned, to fling off a chunk of the planet to form the Moon. The idea remained unchallenged until the early twentieth century, when a number of planetary scientists began to look at the dynamics of the scenario. As with the formation of the whole Solar System, the key factor was the angular momentum, the quantity of the rotational energy stored in the Earth and the Moon.

This is a conserved quantity, meaning that it can never change unless there is an interaction with a third celestial object. So, if you add up the angular momentum of today's Earth and Moon, that must be what Earth alone had, before it spat out the Moon.

If the Moon's angular momentum were placed in the Earth, it would boost the rotational speed, shortening the length of the day to about four hours. As fast as this sounds, it is just not fast enough for Darwin's suggestion to work. To eject part of itself into space, the Earth would need to have been spinning at close to its break-up speed, where the outwardly directed centrifugal force can overwhelm the inward-acting gravitational force. To do that, the Earth would have had to be spinning once every two hours – much faster than seems likely from the analysis.

Nevertheless, the Moon rocks, including the Genesis Rock, were pointing to the fact that the Moon had once been part of the Earth. If that were the case, then they needed to identify something to provide the extra kick: a third celestial body, another planet.

Starting in the 1970s, astronomers began using computer simulation in a serious attempt to develop the hypothesis that a Mars-sized body struck the Earth a glancing blow, which threw up a plume of predominantly rocky material that made the Moon. So confident did they grow with this idea that they named the ill-fated planet Theia, after the mythological Greek Titan who gave birth to Selene.

The scenario was certainly capable of getting the dynamics right. Not only that, but it also seemed a possible fit to the compositional data: the similarity with Earth rocks was because much of the Earth had been blasted off into space to form the Moon. The slight differences were 'contamination' from Theia.

It seemed as if the formation of the Moon was understood, but what no one bargained for was the onward march of technology.

It was 2012 and a sample of the Genesis Rock had been taken from the Lunar Sample Laboratory Facility at NASA's

Johnson Space Center, Houston, Texas, by Hejiu Hui, a geologist from the University of Notre Dame, Indiana. He reanalysed it and other Moon rocks using modern equipment and found that they all contained water.

This seemed somewhat strange for a rock that was supposed to have formed in the hellish aftermath of a giant impact. The heat generated in the cataclysm should have melted all the rocks and driven off the water.

It wasn't the only problem that was becoming apparent. Modern instruments known as mass spectrometers were revealing problems with the Moon's complement of isotopes.

These are simply heavier or lighter versions of familiar chemical elements. They are distinguished by the number of particles called neutrons that each element contains in its nucleus. Often they are associated with radioactivity, but this is not always the case; many stable isotopes of common elements exist as well.

The crucial point is that temperature and density determine how easy it is for different isotopes of the same element to take part in reactions. Hence, planets that formed closer to the Sun will display a markedly different isotopic make-up from those that condensed in the colder outer reaches of the young Solar System. The isotopic composition therefore acts as a hallmark for where the planet was minted.

Increasingly precise analysis techniques have now shown that the Apollo lunar samples have oxygen, chromium, potassium and silicon isotope signatures that are indistinguishable from those of the Earth.

In the giant impact models, the Moon is made predominantly of Theia; hence it should be isotopically distinct from the Earth. In the original analyses, using cruder equipment, there had been enough ambiguity in the measurements to allow for this possibility. However, the message was now crystal clear: the lunar rocks are not just similar to Earth's, they are practically identical. The Earth and the Moon were once the same body.

The news set a couple of researchers thinking. Wim van Westrenen holds a university research chair in planetary evolution at VU University Amsterdam, the Netherlands. His mind started working overtime when his collaborator, Rob de Meijer, a nuclear geophysicist at the University of the Western Cape, South Africa, mentioned deep-Earth reactors, or georeactors.

This is a controversial proposition that, deep inside Earth, overlying the planet's core, are concentrations of uranium and plutonium that form natural nuclear reactors. If they exist, each one could be a few hundred kilometres across and generating energy.

Like many of the planets in the Solar System, Earth emits more energy than it receives from the Sun. This surplus powers the magnetic field, volcanoes and earthquakes. By convention, much of this energy is thought to be coming from the natural decay of radioactive elements deep underground. But, if georeactors exist, they, too, will be contributing.

Assuming that these natural nuclear reactors do exist allowed the researchers to imagine an altogether more explosive scenario for the Moon's origin. The pair reasoned that 4.6 billion years ago there was more radioactive material around, because there had been less time for it to decay away in accordance with its half-life.

Whenever a nucleus did split naturally, it would eject a fast-moving neutron that could do one of two things. It could either split apart another atom, creating more neutrons and sparking a chain reaction. Or, if it struck the right nucleus, such as uranium-238, it could be absorbed and this would transform the nucleus into plutonium-239. While U-238 is not easily broken apart, Pu-239 is fragile. In other words, such reactions would stoke these natural reserves with more fissile nuclear material, making a big chain reaction increasingly likely.

In power stations, this is the way that fast breeder reactors work: they build more fuel than they initially consume. Operators monitor the build-up to make sure it stays within safe limits, but in the putative natural reactors there would have been no such safeguards. Nothing would have stopped one of them building up so much plutonium that it became supercritical and exploded.

To launch enough rocks into orbit to make the Moon would have taken an explosion the equivalent of 40 million billion Hiroshima bombs.

Natural nuclear reactors are not as far-fetched as they sound; sixteen are known about beneath the Oklo region of Gabon, Central Africa. All are now inactive and were discovered in 1972 by the French Commissariat à l'énergie atomique (CEA), which mined the region for uranium. Investigating a significant depletion in the ore of the uranium isotope U-235, the CEA determined that the isotope had been processed as if by a nuclear reactor. Further investigation led to the uncovering of all sixteen natural reactors. Each is between 1.5 and 10 metres across, and they were active some time around 2 billion years ago. They probably continued on and off for a few hundred thousand years, kicking out around

100 kilowatts of power until they exhausted their supply of uranium.

The hypothetical deep-Earth reactors are much larger than those found in Oklo, and instead of the nuclear fuel being concentrated in ore, it would be spread throughout the rocks in minerals such as calcium silicate perovskite. But, perhaps most interesting of all, if the georeactors do exist there may be a way to detect them in the near future.

Radioactive decay often releases ghostly particles known as neutrinos (not to be confused with the heavy neutral particles called neutrons). Neutrinos are extremely difficult to detect. They travel at almost the speed of light and interact with hardly anything. Lead shielding with a depth of 1 light year would not even be guaranteed to stop a neutrino, yet detectors have been built.

They rely on the fact that neutrinos are produced in vast numbers. Here on the Earth, around 60 billion neutrinos pass through the top centimetre of your thumb (and every other centimetre cube of you) every second. The detectors capture only the most minuscule fraction of these, but it is enough. The KamLAND particle detector based in Kamioka, Japan, and the Borexino detector at the Gran Sasso National Laboratory near L'Aquila, Italy, are both registering a steady stream of neutrinos from the Earth's deep interior.

At present, it is unclear whether they are coming solely from the natural radioactive decay of the elements, or whether georeactors are enhancing the release. One way to investigate would be to run a network of neutrino detectors across the world in parallel. It would be possible to coordinate their results and build up a map of the radioactive deposits within the Earth.

With this information it should also be possible to estimate the concentrations and their power output. Reactor regions will give out more neutrinos than natural deposits. Comfortingly, even if natural reactors are found to exist, there is no possibility that one could explode today. This is because the relatively short-lived isotope of plutonium has pretty much all decayed away. The same was not true on the early Earth, say van Westrenen and de Meijer, and they point at the Moon as possible evidence.*

Others are not so sure.

Dynamicists have been exploring whether a smaller, faster-impacting object could generate the equivalent of an explosion, throwing the Moon into orbit. Another scenario suggests that two almost equally sized celestial objects collided and merged to become the Earth, forming a ring of material in the process that became the Moon.

Going from a situation where we thought we knew what was going on to one in which all is confusion again has certainly been frustrating, but it is the cycle of science at work. Theories cannot survive if fresh data does not fit. New hypotheses must be found.

Throughout the Solar System as a whole, composition has been a strongly deciding factor in shaping our theories, but as we start looking at systems of planets around other stars, we are again seeing our original neat ideas thrown into disarray.

Within the crucible of our Solar System, gravity has drawn sufficient matter together to form the Sun, the planets, the moons and the various smaller bodies such as the asteroids. Among the planets, one pattern of order seemed so obvious as to scream for an explanation. The planets closest to the

* http://arXiv.org/abs/1001.4243

Sun are decidedly different from those further away. The Solar System splits neatly into two.

The four planets closest to the Sun - Mercury, Venus, Earth and Mars – are all similar in the sense that they are comparatively small and mostly made of rock and metal with relatively tenuous atmospheres. Mercury is the smallest planet at just 4879 kilometres across; Earth is the largest at 12,742 kilometres.

The four are spread across 500 million kilometres of space, with the Sun at the centre of this system. Mars's orbit at 250 million kilometres from the Sun defines the outer boundary. Sun-baked Mercury is tucked just 58 million kilometres away from the Sun and Earth rests at 150 million kilometres, a measure that is known as the astronomical unit.

The asteroid belt comes next, spanning the distance between Mars's orbit and that of Jupiter at 779 million kilometres. There are more than half a million known asteroids in this ring of space. The largest is Ceres, at 950 kilometres in diameter. The smallest is about 5 metres, but this lower limit is constantly being extended as better telescopes come online.

The asteroids represent the leftovers of planet formation. As we saw in the last chapter, the whole Solar System was once a giant asteroid belt, with billions upon billions of asteroids, called planetesimals, that merged to produce the planets we see today.

The asteroid belt was prevented from coalescing into a planet because of the gravity of Jupiter. Sitting at the outer edge of the belt, the giant planet's gravity has corralled the asteroids into more or less orderly orbits, preventing them from touching and growing into a fifth rocky planet.

Jupiter has such influence because it is huge: the largest

planet in the Solar System. Along with Saturn, Uranus and Neptune, it belongs to the second group of planets known as the gas giants. These are all relatively large, with deep atmospheres surrounding a core of rock and metal.

The scale of these planets is staggering. Jupiter is just shy of 140,000 kilometres across – almost eleven times the diameter of Earth. It is enshrouded in an atmosphere that masks the interior from view and contains a hurricane-like storm, known as the Great Red Spot, so large that it would engulf our entire planet.

Saturn, with its ring system of rocks and pebbles like a mini-asteroid belt all of its own, is almost the same size. Uranus and Neptune are smaller, but still they are around five times the size of the Earth. Neptune lies at 4.5 billion kilometres from the Sun and defines the outer edge of the known planetary system.

There is an appealing symmetry to the arrangement: four terrestrial planets in the inner solar system and four gas giants in the outer. It cannot help but trigger the feeling that such order must surely imply some underlying principle at work, and indeed, the nebular hypothesis offers a ready solution. With the young Sun in the centre of the forming Solar System, those parts of the nebula closest to the Sun will have been heated more than those far away. The inner parts of the nebula are heated so much that only the so-called refractory elements, which are resistant to heat, can become solid. These are mainly rocks and metals. The volatile material that needs cold temperatures before it can become solid in the form of ice, did not stand a chance. Hence, close to the Sun, planets rich in rock and metal formed: Mercury, Venus, Earth and Mars.

Further away, the temperature became more moderate and at about five times further from the Sun than the Earth, ice began to crystallize. It was not just made from water but also carbon dioxide – so-called dry ice – ammonia and methane. Together these are referred to as the astronomical ices.

These boosted the amount of solid material available to the forming planets in this region of the Solar System, allowing them to become naturally larger. Each planet grew by feeding from the planetesimals in its orbit. The further a planet is from the Sun, the larger the circumference of its orbit. So, again, the outer planets had more material from which to grow.

These two factors combined to make the outer planets larger than the inner ones. Then gravity really kicked in because once the outer planets had grown sufficiently large, they generated a gravitational field big enough to pull in gases from the nebula.

Gases are the flightiest atoms of them all. It takes a strong gravitational field to hold on to a gas, even one of the heavier ones like oxygen. This is why the Moon, with just one-sixth the gravity of the Earth, has no air. It simply cannot muster the gravitational strength to hold on to the stuff. Earth can retain a thin layer, just more than 100 kilometres in depth, consisting mainly of nitrogen, oxygen and carbon dioxide. Lighter gases such as hydrogen and helium are virtually absent from our atmosphere.

The nascent planets of the outer Solar System, in particular Jupiter and Saturn, grew so large that their gravitational fields could retain any gas, including hydrogen and helium, the very lightest elements of all. The only other celestial body capable of doing this was the Sun.

So today, Jupiter and Saturn have compositions that mirror the Sun's, yet neither became large enough to set off the necessary energy-generating reactions deep inside to turn them into stars, as we shall see in the next chapter.

In this way, explaining the architecture of the Solar System seemed so obvious. It was glorious natural order springing from nothing more than simple physical laws. Astronomers expected fully that when they developed telescopes capable of detecting planets around other stars, the arrangement of planets in our Solar System would be repeated.

In 1995, the dream came true. The first 'exoplanet', the term for such a world around another star, was found. But the dream quickly turned into a nightmare because nothing was as expected.

The discovery of the first planet around a Sun-like star fell to a pair of Swiss astronomers, Michel Mayor and Didier Queloz. Based at the Geneva Observatory, they were using a telescope in the Haute-Provence region of France to look for planets. A novel approach was required because planets are so dim compared to their parent stars.

It is almost impossible, even today, for a telescope on Earth to distinguish a planet from the glare of its star. So Mayor and Queloz looked at the light from the star to see if it was 'wobbling'.

We talk about planets orbiting stars as if the star is the stationary centre of the system, the hub around which everything turns, but in reality that's not quite true. The orbiting planet generates its own gravity and so pulls back on the star. This force is not enough to swing the star into a large orbit because the planet is much less massive than the star, but it can force the star to pirouette, a motion that corresponds to a small orbit. This means that the star is moving and so sometimes it will be heading towards us and sometimes away from us. This minutely affects the quality of the light emitted by the star. Whenever the star is travelling towards us, the light is 'squashed' and its wavelength shortened. On the opposite side of its pirouette, the star will be moving away from us and its light will be stretched to slightly longer wavelengths.

This is a manifestation of the Doppler effect, one of the most powerful tools in the astronomer's arsenal for diagnosing the celestial realms simply because it betrays a celestial object's movement towards or away from us.

The time it takes the star to complete a pirouette is revealed by the sequence of stretching and squashing in the light, and is equivalent to the time it takes for the planet to traverse its orbit. Once the planet's orbital period is known, the diameter of its orbit can then be calculated, so the planet's distance from its star is revealed. Combining this with the magnitude of the squashing and stretching in the star's light then gives the planet's mass.

In 1995, Mayor and Queloz had a puzzling signal. Applying the analysis suggested that they had found a planet around the star 51 Pegasi. But in comparison to our own Solar System, nothing made sense. The planet was a gas giant at least half the mass of Jupiter, yet it was far closer to its star than Mercury, which completes an orbit in 88 days. 51 Peg b, as the new planet was termed, shot round its orbit in just 4 days. It is so close to the star that its atmosphere must be a searing 1000 °C. Since the heat a planet receives drives its storms, 51 Peg b's atmosphere must be a maelstrom.

All in all, this newfound world seemed an unreal place –

so unreal, in fact, that doubts were raised about its very existence.

All uncertainty was swept away within weeks, when an independent American group analysed data they had been sitting on. 51 Peg b was found there too. More than that, they announced other planets around other stars. Gas giants could indeed be found in close proximity to their parent stars. In one fell swoop, astronomers had found not just a new planet, but a new class of planet, and swept away the belief that gas giants could only be found at large distances from their host stars.

According to the theories of planet formation, designed with our Solar System in mind, this was impossible. There simply was not enough matter that close to a star to build a gas giant. So, said the theorists, 51 Peg b and its ilk must have formed far from the star and then migrated. The theoreticians set to work on wondering how planets could move about and found it easier than they expected.

A planet moving through its orbit feels no resistance because there is no air to drag on it, so in principle a planet could circulate indefinitely. Back during the birth of the planets this was not quite true. The protoplanet was ploughing through the remains of the cloud that collapsed to form the star. This could have provided a drag that set a large planet spiralling inwards towards the star.

En route, its mighty gravity would have scattered the interior planets into wild orbits or even pushed them to a fiery death in the central star. The more the theoreticians have looked, the more possibilities they have found for this kind of behaviour.

At the same time, the more observers have looked, the more types of planets and arrangements they have found.

There are now more than 1000 identified exoplanets. Nearly two hundred multiple planet systems have been revealed, and one thing is becoming very clear: our Solar System with its orderly collection of four terrestrial planets and four gas giants is not normal – far from it. Instead of being the norm, it may prove to be the exception to the rule.

The only way we will know is for astronomers to continue developing more and more sensitive spacecraft for detecting smaller and smaller planets. There are a number of missions being readied for launch that will do this. NASA is building the Transiting Exoplanet Survey Satellite (TESS) for launch in 2017; ESA is working on the CHaracterising ExOPlanets Satellite (CHEOPS) also for 2017, and the PLAnetary Transits and Oscillation of stars (PLATO) for 2024. These will reveal whole solar systems, not just the larger planets. It will truly allow us to see if any patterns emerge or whether we are just living in an oddball system of planets that just happens to be nicely arranged.

But, before astronomers get completely carried away finding planets around other stars, it is perfectly possible that there are additional planets around our own Sun waiting to be found. Maybe there is even a whole new solar system out there in the icy depths beyond Pluto.

My phone rang as one day was about to turn into another. It was the last moments of Friday, 29 July 2005, and my first thought was to ignore the call. It could not be a family emergency because it was the office phone that was ringing. I had well and truly finished work for the week, but something made me pick it up.

'Hello?' I deliberately sounded a bit vague.

It was Steven Young, the publisher of *Astronomy Now* magazine. I was the editor of the magazine at the time and Steve lived and worked in Florida, so I was used to him being in a different time zone, but even so this was unusual.

'It's midnight,' I said.

'I know. But NASA is about to announce the discovery of Planet 10.'

Since the discovery of a new planet in our Solar System had not happened since 1930, I was prepared to work some serious overtime.

Minutes later, I was hooked into a hastily arranged press briefing in which Mike Brown of the Californian Institute of Technology described the discovery of a distant world, larger than Pluto. My grip tightened on the telephone. The implication was clear: if Pluto was a planet, so was this.

2003 UB313 was the official designation, but as the press quickly picked up on, Brown and his colleagues, Chad Trujillo and David Rabinowitz, jokingly refer to it as Xena, after the heroine in the television series *Xena: Warrior Princess*.

Almost instantly, the name of the planet became the central story. As discoverers, Brown, Trujillo and Rabinowitz had the privilege of choosing it. Brown had originally wanted to call the world Lila, a concept taken from Hinduism that can be used to describe the cosmos as the outcome of the divine being's pastime or sport. It was also close to the name of his newly born daughter.

But before any naming could take place, the International Astronomical Union decided it was finally time to define what was and wasn't a planet. Until this point, there had been no definition. Most of the planets in our Solar System were known since antiquity. They were the 'wandering stars' visible to the naked eye that we discussed in Chapter 1. On top of these, only three more planets had ever been discovered in orbit around the Sun.

The first, Uranus, was spotted by William Herschel from his back garden at 19 New King Street, Bath, England, on 13 March 1781. The second, Neptune, was bagged by Johann Gottfried Galle and his student Heinrich d'Arrest of the Berlin Observatory on 23 September 1846. In Herschel's case he was conducting a systematic search of the night sky. Galle, on the other hand, was working from a predicted position of the planet given to him by French mathematician Urbain Le Verrier.

Then, in 1930, it was the turn of Clyde Tombaugh from the Lowell Observatory, Arizona. On 18 February, he found the planet Pluto as he compared photographic plates of the same region of the night sky. He noticed that in the six days between the photographs, one object in the field of view had moved. The discovery made headlines across the world, but it soon became obvious that Pluto was not as mighty as the other planets.

Follow-up observations showed it to be a minnow. It has a diameter of just 2300 kilometres, making it smaller than our Moon. Its volume is less than 1 per cent of the Earth's, meaning that it has a total surface area equivalent to that of Russia. Could such a comparatively small object really be called a planet?

The question was moot until the discovery of 2003 UB313. Now, a decision had to be made because astronomers were on the hunt for more of these objects. The International Astronomical Union set up a committee of seven individuals under the chairmanship of Owen Gingerich, a Harvard

University historian of astronomy. The committee also included the best-selling science writer Dava Sobel. Together, they drafted a proposal for 'planethood'.

In August 2006, the International Astronomical Union met in Prague for its biannual general assembly. It was a fitting place, as this was where Johannes Kepler first met Tycho Brahe and began the work that led to the mathematical description of planetary orbits.

The committee's draft proposal was put before the membership. It said that two conditions had to be met for a celestial object to be deemed a planet. First, it had to be in orbit around a star, while not itself being a star. This was not contentious. Stars, as we will see in the next chapter, must be large enough to generate their own energy by means of nuclear fusion reactions. This is what makes them shine. Planets, however, appear bright because they are reflecting starlight.

The second condition was that a planet must generate enough gravity to be capable of pulling itself into an almost spherical shape. This would be the case for an object with a mass of around 500 billion billion kilograms, and a diameter of about 800 kilometres. In this way, the committee hoped that they were using the strength of gravity, hence the natural laws of the Universe, to underpin the definition, but there was a price to pay.

It would require the reclassification of several Solar System bodies. The largest asteroid, Ceres, would now be a planet. Pluto's moon, Charon, would also be a planet, making the two a double-planet system. Together with the nine traditional planets and 2003 UB313, that would mean the solar system contained at least twelve planets. The committee pointed out that astronomers knew of a dozen other asteroids that further observations could prove were also planets by this definition.

It proved too much for many astronomers because it required too much upheaval in the way we thought about the Solar System. Just two days after the draft resolution was published, amendments started to be proposed by the membership. Eventually, after heated debate and some acrimony, things came down to the wire. It was the last day of the conference before the IAU executive put forward the amended definition for vote.

A crucial additional clause was added. To be considered a planet, the object must have 'cleared the neighbourhood around its orbit'. This would rule out the smaller bodies such as Pluto because they share similar orbits with other celestial objects. Pluto, for example, is the largest of the Plutinos, a group of asteroid-like objects that have been corralled into similar orbits by Neptune's gravity.

It would mean that the Solar System consisted of just eight planets. Pluto would be part of the newly defined class of dwarf planet, along with 2003 UB313 and a retinue of other distant little worlds.

Of the 2700 astronomers that had attended the general assembly that year, fewer than 500 were still there on the last day. Nevertheless, the vote went ahead and revealed there was overwhelming support for the amended definition. The die was cast, and Pluto was demoted from planet to dwarf planet.

Instead of becoming only the fourth man in history to discover a planet, Mike Brown became the first to cause the demotion of one. He embraced the infamy, calling himself @plutokiller on Twitter, and writing a book titled *How I Killed Pluto and Why It Had It Coming*.

When it came to the official name for 2003 UB313, Brown

and his colleagues proposed Eris, the Greek god of chaos, strife and discord. It was accepted – rather appropriately, given the dispute its discovery had sparked.

For now, the controversy has quieted down, but there are reasons to believe that a whole host of additional bona-fide planets are orbiting the Sun at extreme distance and that the next generation of survey telescopes will find them.

It comes down to the way planets were built in the series of collisions between smaller objects called planetesimals as discussed in chapter 1. To simulate this process, astronomers develop computer models that track the growth of the planets. These show that in the inner Solar System there were probably twenty to thirty Mars-sized planets that grew into the four terrestrial planets Mercury, Venus, Earth and Mars.

In the outer solar system, the same simulations show that there were probably a similar number of planets, but this time about the size of the Earth. These coalesced into the cores of the giant planets Jupiter, Saturn, Uranus and Neptune.

In other words, about 4.5 billion years ago the Solar System contained around forty to sixty rocky planets that fought it out for supremacy, ingesting their rivals and growing into the planets we see today. With this dog-eat-dog scenario in mind, astronomers have taken to referring to the situation as the oligarchic scenario.

Computer simulations by Scott Kenyon at the Harvard Observatory show how it would have taken place. Initially, the oligarchs kept to their orbits, circling the Sun as if sizing each other up. Periodically they drew close to each other and their pull of gravity caused subtle changes in the shape and size of their orbits. These changes built up until, suddenly, things went out of control.

A chaotic period ensued in which oligarchs were flung all over the Solar System. Many collided and coalesced to become the familiar planets of today, but not all. Near misses were as likely as collisions, and according to the computer simulations, a near miss between an oligarch and a burgeoning gas giant planet could be enough to throw the smaller object into exile.

Some would be given so much of a kick that they would have broken free of the Sun's gravity. Even now they will be wandering the Galaxy as frozen planetary orphans. Most of the exiles won't have made it that far. They will remain in the Sun's gravitational grip forming a 'cloud' of planets that loop the Sun in giant orbits taking between tens of thousands and millions of years to complete. They will be found in orbits 25 to 250 times further from the Sun than Pluto, and because the gravitational kick given to these oligarchs alters the size, shape and orientation of their orbits, they will be scattered at random angles to the original disc containing the rest of the planets.

It is impossible to know the precise numbers of worlds that might be lurking out there, but Kenyon's simulations suggest that between 10 and 20 per cent of the oligarchs suffered this fate, meaning that between six and twelve Earth- to Mars-sized planets are waiting to be found. If true, it is the equivalent of a whole new solar system's worth of planets.

There could well be circumstantial evidence for this population already. Unlike the rest of the planets, which spin more or less at right angles to their orbit, the seventh planet, Uranus, spins on its side. This gives it an extreme version of the Earth's midnight Sun or polar night phenomena. On Uranus, the night can last for decades, followed by decades of daylight, as the planet crawls around its 84-year orbit. But what could tip a planet on its side like this? Perhaps it was a collision between the gas giant and a speeding oligarch on its way out of the Solar System.

The only way to be certain whether these extra planets are really there is to see them. Despite their size, that is a tough undertaking because they are going to be very faint indeed. The sunlight needed to make them visible will be feeble at such distance.

Compounding the problem is the fact that the gravitational scattering which sent the planets out there in the first place is a chaotic process. This means it robs astronomers of any way to predict where the planets are likely to be hiding. The only way to search for them is to trawl the whole sky using telescopes, and this venture is so time-consuming that it is way beyond the scope of most observatories, which usually allocate time to astronomers in chunks of just two or three nights.

But the search may be receiving a boost from a new project called the Large Synoptic Survey Telescope. To be sited on Cerro Pachón, Chile, LSST is scheduled to begin work in 2024. With a mirror measuring 8.4 metres across, it will be roughly ten times as sensitive as anything available today for such surveys. Calculations show that it could detect an Earth-size planet up to 500 times further away than our own planet is from the Sun.

The telescope will be a monster that devours the sky. Its wide-angle camera can survey the entire visible sky in just three days, and then go back and start again. Every night it will generate 30 terabytes of data: enough to fill the hard drives of several hundred PCs. Automated software will detect moving celestial objects in this ocean of data and post the coordinates on a public website for others to follow up.

Although professional observatories will be working hard to stay on top of the follow-up observations, the public nature of the data opens the door for amateur astronomers to discover a new planet in the Solar System. Would-be planet spotters should keep their eyes peeled for unusually slow-moving, faint objects in the data, preferably far away from the orbital disc of the Solar System.

So, if you fancy going down in history, all you need to do is wait until LSST fires up, then get ready with a telescope, an Internet link, and an awful lot of patience and good luck.

Yet no matter how many new planets may or may not be awaiting discovery in those far-flung realms, it is in the centre of the Solar System that we find gravity's true crucible. This is the Sun.

In terms of matter content, the planets are mere trivialities. The Sun contains 99.8 per cent of all the matter in the Solar System. All eight planets, the dozen or so dwarf planets, the many dozen moons, and the billions of asteroids and comets: put all of these together and you still only reach 0.2 per cent of the total mass in the Solar System.

Since mass is all that really matters when it comes to generating gravity, the Sun is by far the gravitational hub of the Solar System. Its gravity extends into the far reaches of space, shepherding comets out to 50,000 times Earth's distance from the Sun, and it contributes to the overall gravitational field of the Galaxy, together with the other 100 billion or so stars that make up the Milky Way.

Gravity works inside the Sun as well. The weight of the overlying layers bears down on the core of the Sun, creating a pressure more than 250 billion times that experienced in

Earth's atmosphere. This ratchets up the temperature to more than 15 million °C.

Under these extreme conditions, new atoms are forged from old. Hydrogen is fused in a series of nuclear reactions to become helium. In the process, energy is released, and this is the principal definition of a star: that it shines by making its own energy. To keep the Sun shining, it must convert about 4.2 million tonnes of matter into energy every second.

This energy is what becomes sunlight but it takes hundreds of thousands of years for it to fight its way out through the crushing weight of the surrounding gas to reach the surface of the Sun 695,000 kilometres overhead. Once there, the light speeds off into space, crossing the 93 million miles to Earth in just eight minutes.

The extraordinary truth is that the sunlight rising from the eastern horizon every morning was generated inside our luminary before any human walked the Earth. But to deduce this required a leap of faith in the way that science is practised, and opened the door to the models that would rise to dominate cosmology.

3
Gravity's Crucible

The Sun is the central engine of the Solar System. It is the crucible into which nature has drawn 2×10^{30} kilograms of matter. Not only does it produce the gravity that keeps Earth and the other planets in orbit, it also generates the heat and light that keep our world alive. No other celestial object has been studied so often or in so many different ways, yet only reluctantly has it given up its secrets.

This is because all of the true action takes place deep in the Sun's core, buried away from view. The emission at the surface is merely the waste product of these hidden reactions. And at the beginning of the twentieth century that was putting astronomers in a bit of a fix.

There was no known energy source powerful enough to produce the Sun's luminosity. Every second, the Sun shines 3.85×10^{26} joules of energy into space, which is roughly enough every second to keep humans in power for 100 million years. Chemical reactions and even radioactivity are simply incapable of sustaining the Sun's prodigious welter of energy.

The only process that seemed to offer any hope was gravitational contraction. First proposed in the nineteenth century as the Sun's energy-generating mechanism, it worked very simply. When a gas is compressed, its pressure rises and so does its temperature. In the case of the Sun and the other stars, gravity was pulling them together, thus compressing the innards. So gravitational energy, called potential energy, was being converted into heat, which was then radiated from the star.

The sticking point was that the maximum amount of energy that could be generated in this way would only sustain the Sun's furious output for a few million years, and there was a growing handful of dating techniques, involving the newly discovered radioactive elements, that suggested Earth's age should be measured in billions of years rather than mere millions.

Then, in 1905, Cambridge astrophysicist Arthur Eddington read something that changed science forever. It was a paper by an unknown German-born physicist by the name of Albert Einstein, who was working at the Swiss Patent Office in Berne.

The paper was published in November of that year. It did not relate to any specific observations but it put forward a speculation that the inertia of a body (its tendency to resist movement) could change depending upon the amount of energy it either absorbed or emitted. If a body absorbed energy, then it would become heavier, and vice versa if it lost energy. It was the work that led to the world's most famous physics equation: $E = mc^2$.*

In the equation, E equals energy, m equals mass, and c is the speed of light. It shows that matter and energy are interchangeable; light can be transformed into particles and vice versa. It is a behaviour that is hardwired into the physics of the Universe. Amazingly, Isaac Newton seemed to presage

* The famous simplified form of the equation did not emerge until 1946, when Einstein used it in the title of a paper.

the discovery in 1717. In the fourth edition of his book *Opticks*, he wrote: 'Are not the gross bodies and light convertible into one another, and may not bodies receive much of their activity from the particles of light which enter their composition?'

Einstein encapsulated this conjecture mathematically, giving the precise amount by which the mass of an object goes down when it emits radiation and up when it absorbs. Although experimental confirmation did not come quickly – it took until the 1930s to be able to test the $E = mc^2$ concept with laboratory equipment – for Eddington, Einstein's original paper was enough. It set his mind racing because he saw it as a way to make the Sun shine.

It suggested that mass could be converted into energy and radiated, and if there was something that the Sun had in vast abundance, it was mass. But for Eddington to get his fellow scientists to accept his ideas as anything more than idle speculation meant changing the way science was performed. To do that, Eddington had to go to war with one of the leading physicists of the day.

James Jeans cut a formidable figure. His portraits show a confident individual who looks squarely into the camera lens. Of substantial build, with a neat sideparting, the downturn to his mouth confers an air of gravitas and contemplation. He was a talented mathematician who believed in the old-school approach that science should be built from the bottom up, brick by brick. Each new advance should rest on top of an incontrovertible mathematical or observational proof.

This had been the way of science since Newton's time. While it clearly led to a solid chain of proof, it restricted the problems that could be tackled. Without the bedrock of measurement, Jeans argued that nothing could be calculated with certainty – and therefore it wasn't science.

He singled out astronomy in an article he wrote in 1909, stating that mathematicians wanting to pursue astronomy found themselves in a difficult position. Without a way to get to the stars and measure their interiors, how could you ever be certain of their internal conditions? According to Jeans, astronomical analyses that proceeded from a sound basis of assured physical facts were only really possible for those studies using Newton's laws of gravity. In other words, you could analyse the motion of the stars and planets but knowledge of their interiors was impossible. For Jeans, this was the necessary cost of keeping science pure. But what good was such purity to astronomers hungry to investigate the Universe?

In 1910, Eddington was commissioned by the *Encyclopaedia Britannica* to write the entry for 'star'. His description contained the admission that stars 'might be solid, liquid, or a not too rare gas'. Since no other states of matter were known to exist at the time beyond solid, liquid and gas, it was the equivalent of saying that astronomers simply didn't have a clue. Surely, he thought, any insight that could be brought to bear on the subject would be preferable to this void.

Unlike Jeans, Eddington had not received his training solely in the field of mathematics. Before reading mathematics at Cambridge, he had received a physics degree from Manchester. After Cambridge, he worked for seven years as an astronomer in the Royal Greenwich Observatory. With this background, he was primed to assimilate all of the discoveries in physics that were taking place at this time, and use them to guide a mathematical understanding of the stars.

In his mind's eye, he nurtured the inspiration that the Sun and the other stars were powerful engines generating energy via $E = mc^2$. All he had to do was find a physical process that satisfied the equation. As it turned out, he didn't need to look far at all because physics was alive with the discovery of radioactivity.

In Paris, Marie and Pierre Curie were pioneering the investigation of these strange particle-spitting atoms. In Cambridge, Ernest Rutherford was on the path to splitting the atom. Both investigations were pointing to the atom as a source of great energy, one governed by Einstein's equation and easily enough to power a star.

But how could Eddington prove that atoms were the source of stellar power? Jeans was right, there was no direct measurement of the Sun's core, no direct approach to solving the problem that he could use.

Eddington himself was not the most imposing of figures. Surviving portraits show him as anxious in front of a camera; the expression on his thin lips could be mistaken for a sneer. But when it came to his intellectual pursuits, he was a hammer.

Stepping away from the pure tread of the scientific method, Eddington wondered if he could reverse-engineer a solution – in other words, invent a situation based on what sounded like reasonable physics for a star, and use the resulting mathematics to calculate observable outcomes. For example, if he just assumed that a star was made of gas, he could use nineteenth-century physics to calculate quantities such as its central temperature.

He set to work, and on 8 December 1916 he stood before the gathered fellows of the Royal Astronomical Society and explained what he had found. He acknowledged that the conclusions his method gave were tentative, and may never be proved. The analysis, he admitted, could only be based upon the most general laws of nature. Nevertheless, he argued strongly that such an approach could yield insights into stars if they really were gaseous objects.

Take, for example, a discussion of how radiation flows through the star to the outside Universe. No one could doubt this happened since the stars were shining, but a mathematician would demand to know the composition of the star so that its opacity could be calculated. This is the figure that quantifies how easily radiation can travel through a medium.

It was impossible to get inside a star and take this measurement, but an astronomer could measure the energy shining from its surface and use that to back-calculate something about the opacity. He could then tinker with the possible compositions until finding one that fitted the inferred opacity.

Yes, it was turning the method of inquiry on its head and, in doing so, splitting astronomy away from mathematics and its methods. Yes, it was a dangerous path to tread because now you could never be absolutely certain about your conclusions; at best, you could say that they were probable, but the alternative was simply to abandon the attempt to understand the stars.

What Eddington was proposing wasn't a theory of a star's interior: it was a model. The key difference between a theory and a model is that a theory can tell you *why* something happens but a model can only describe how something *may be happening*.

So long as the model's results were found to be in accord with observations that could be made, such as the brightness of the Sun, then all was well. Eddington thought that the end justified the means. Models could always be tweaked, substantially revised, or thrown away and replaced as new observations became available, but in lieu of direct observations, they offered *the only way* to make progress.

As expected, Jeans attacked at once, arguing that this is what science had been devised to avoid: belief without proof. Central to his assault was the lack of definite knowledge about how the Sun generated energy. Jeans thought that this was the foundation stone on which to base any treatment of stellar structure. Eddington may have been willing to speculate that atomic power was the answer, but without a full mathematical treatment of this process, Jeans simply behaved as if it did not even exist.

Eddington battled on. It was long, arduous work, and only in 1924 did it result in a clear intellectual bridgehead. He showed that, by assuming a star was made from gas, the energy it radiated every second was directly proportional to its mass. The more massive the star, the hotter is its surface and the greater its luminosity.

Eddington's model could not tell him precisely why this should be so but it was in good agreement with observations. Buoyed by such progress, he summarized his work in a landmark book called *The Internal Constitution of the Stars*. Publishing in 1926, and ever mindful of his fight with Jeans, he was at pains to acknowledge the differences between his and the mathematician's way of doing things. Indeed, the treatise was as much about the new method of modelling as it was about the stars themselves. By talking about the latter, he described the former.

He admitted that it might be impossible to know exactly

how a star worked, yet it was surely within our grasp to understand something about which physical processes were important and which were inconsequential.

The downside to all of this, as Jeans rightly feared, was that it opened the door to error. There was always the possibility of drawing the wrong conclusion. Take, for example, a person who goes to the horse races rather infrequently. As luck would have it, every time he has been in attendance, a jockey wearing a red cap has won the race. Based on this evidence, he could draw the conclusion that all races were won by the jockey wearing the red cap.

Clearly this is rubbish: our common sense tells us that the colour of the cap has no bearing on the performance of the horse or its rider. But in the world of astronomy, where most things are way outside human experience, Eddington warned that modelling could lead to disastrous consequences if misused: wrong conclusions, total blind alleys, years of wasted effort. But it was a necessary risk, he argued, and could be mitigated by the use of physical insight, the scientist's equivalent of common sense.

And so the use of models to deduce physical parameters was born into astronomy. The resulting models were only ever probable and, in this way, astronomers may hope to approach the truth, even if the certitude of mathematical proof may always elude them.

In the final chapter of *The Internal Constitution of the Stars*, Eddington returned to his intuition about $E = mc^2$ and speculated about the release of atomic energy in stars.

There was mounting evidence from the various laboratories across Europe that atoms could be transformed into one another, either by being split or by being forced together. The one that captured Eddington's eye was the fusion of hydrogen into helium.

At the heart of each hydrogen atom is a single particle known as a proton, making it the lightest chemical element. The next heaviest element is helium, with a nucleus approximately four times heavier. This made it look inevitable that helium was the combination of four hydrogen nuclei, but measurements were showing that there was a discrepancy in the mass. A helium nucleus was about 0.8 per cent lighter than the four hydrogen nuclei that went into its composition. Eddington advanced the hypothesis that this 'mass defect' was converted into energy according to Einstein's equation and radiated into space, giving the stars their shine.

The trouble was that the temperature needed to force the hydrogen nuclei together was calculated to be huge: about 15 billion °C. Although Eddington's model was pushing in the right direction, it fell far short with an estimated central temperature of the Sun at just 40 million °C.

This invited criticism both of his model and of him. Rather than give in to his detractors and their calls for him to abandon his ideas, he became more intransigent than ever. He fought back in the book with a masterstroke born of complete self-confidence: 'But the helium that we handle must have been put together at some time and some place. We do not argue with the critic who urges that the stars are not hot enough for this process; we tell him to go and find a hotter place.'

By this he meant put up or shut up. Maybe Eddington's stellar model was not quite right, but until something came along that was demonstrably better, it would do. This was conviction science, and a perfect example of the new way of advancing astronomy. Even if the model seemed highly improbable, it remained the best until something better came along.

As it turned out, nothing better was needed because the breakthrough Eddington needed came the very same year that he published his book. German theoretician Werner Heisenberg was working at the Institute for Theoretical Physics at the University of Copenhagen, alongside the acclaimed physicist Niels Bohr.

With his long face, fleshy lips and passion for football, Bohr had won the 1922 Nobel Prize in Physics for his description of atomic structure. Now known as the Bohr model, it is the one taught in every classroom and depicts the particles as miniature billiard balls in motion around each other. The Bohr atom contains a small, positively charged nucleus surrounded by negatively charged electrons that travel in orbits. It is somewhat similar to the behaviour of planets around the Sun except it is the force of electromagnetism, with its positive and negative charges, moving the particles rather than gravity.

Bohr developed the model firstly to explain hydrogen, the simplest of all the atoms, with its nucleus made of a single proton. This attracts a single, negatively charged electron, which goes into orbit making the atom electrically neutral.

The electron can inhabit one of a number of different orbits and can jump between them by absorbing or emitting energy. Its chosen diet is light, but not just any old ray. The light has to carry the exact amount of energy that will move the electron into a higher orbit. When the electron falls down to a lower orbit, it re-emits this light. It is this movement from one energy state to another that is known as a quantum leap.

Bohr showed that different elements possessed different sequences of electron orbits. Different elements could be described by different numbers of protons in the nucleus and different numbers of electrons in orbit around this, and it was the differing arrangement of the electron orbits and the number of electrons in the outermost part of the atom that give rise to the chemical properties of each element.

With this as a base, Heisenberg was trying to work out the details of how an electron orbits a nucleus, but the mathematics was refusing to give a precise trajectory. Instead it seemed to imply that one can never know with perfect accuracy the two most important factors in pinpointing the electron: its position and its velocity. The more precisely one quantity is known, the less precise it forces the other to be.

This is now known as Heisenberg's uncertainty principle and is a cornerstone of quantum mechanics, which aims to understand the motion of nature's smallest objects: the atoms and the subatomic particles.

On everyday scales, the position and velocity of objects can be precisely known. Just ask anyone whom the police have stopped for speeding. These are based on the laws of motion given by Isaac Newton. But in the realm of particles, Newtonian mechanics no longer apply. Instead, different rules take precedence; these are known as quantum mechanics.

At first glance, Heisenberg's uncertainty principle looks as if it is telling you something about the precision with which you can make measurement. In other words, while you may not be able to measure the position and velocity of a particle precisely, surely the particle 'knows' where it is; otherwise, how would it know when it comes into contact with other particles and interacts with them? Amazingly, Heisenberg's uncertainty principle is not telling us anything about measurement; it is telling us about reality. The limits of precision with which the particle's position and momentum can be known are hard-wired into the Universe and that means the particle is as confused as everybody else about exactly where it is.

It solves Eddington's mystery about how the Sun generates energy because it means that the particles are not solid little cannonballs as Bohr pictured them but much larger fuzzy regions of inexactitude. The size of this region is given by the Heisenberg uncertainty principle. If these regions touch, then it is possible for the two particles to interact. The more they overlap, the more probable the interaction becomes. Nature, it seems, does not deal in the black and white of hit or miss.

In the original fusion calculations, physicists had assumed that the protons were acting like billiard balls and needed to collide. Hence, they calculated the tremendous energy needed to force them together. In reality the particles did not have to touch, just get close enough for their clouds of uncertainty to overlap. The phenomenon is called quantum tunnelling.

While it raises profound questions about the nature of reality, which we will look at in our final chapter, Eddington could hardly have cared less because the practical upshot was that this extraordinary piece of weirdness significantly lowered the temperature at which fusion could take place. The particles no longer needed to be forced together, just forced close to one another. Indeed, as the numbers unfolded in front of him, Eddington saw that they were now in line with the temperatures he was estimating from his model of the Sun's interior. The vindication he must have felt!

Surely such agreement was too unlikely to be mere coincidence. It must be pointing to the fundamental truth of

Eddington's model. If so, it was a spectacular justification of this approach and showed that a compelling explanation could be constructed by reverse-engineering the problem.

From that day to this, astronomers have believed that the Sun and stars are nuclear furnaces, most of them shining brightly because they are fusing hydrogen into helium.

In the century since Eddington first suggested modelling the Sun, observations have become more and more accurate, computers have become more and more powerful, and models have become more and more detailed. Yet despite this, the Sun still has the ability to take us completely by surprise – and in doing so, reveal the gaps in our knowledge about its inner workings.

The only visible marks on the solar surface are the sunspots. Sometimes there are a multitude of these dark blemishes; at other times none at all. It all seemed random and unpredictable until a disillusioned German chemist named Heinrich Schwabe sold the family apothecary shop in the early 1820s and spent the proceeds on a very fine telescope indeed. His intention was to indulge his intellectual passion for astronomy rather than scrape a living making medicines and ointments.

Canvassing university professors for a project worthy of such an optical instrument, he was told that a systematic study of the sunspots would be useful. No one had attempted it before and there were two sweeteners for his efforts. First, and by contrast with the majority of astronomical investigations, no observations would be required at night, so he didn't need to lose any sleep. Second, he might just spot an unknown planet crossing the face of the Sun, and therefore become only the second man in history (after William Herschel) to discover a whole new world. Put like that, how could Schwabe resist?

He set up the telescope in his loft, pointed it at the Sun, but instead of looking at the burning disc through the eye-piece,* he projected the resultant image onto a screen. Then every clear day he recorded the number of sunspots.

He stuck at these observations for years and years. He was diligent and exacting in the face of what must have been utterly repetitive work. Clearly, some inner drive kept him returning to the attic and his patience was finally rewarded about fifteen years into the endeavour. Although he didn't spot a planet, he did notice a pattern in the annual number of sunspots.

Some years there were hardly any at all. Gradually the number increased year after year, until it reached a peak. Then, more quickly, the number dropped to almost zero again, and the pattern began to repeat. All told, the cycle took about eleven years to complete. This was the first nugget of information about the unknown generator deep inside the Sun that was producing the sunspots.

The second was that the sunspots were magnetic. In the years of their greatest number, compasses on Earth deviated significantly from magnetic north, forcing mariners to make corrections in their navigation. As impossible as it sounded, somehow the sunspots were reaching out across 93 million miles of void and deflecting the magnetic needles on Earth. This was the first piece of evidence that a force other than gravity could transmit itself across space. Until then, magnetism and electricity had been assumed to be short-range

* Never look directly at the Sun with or without a telescope. It is so bright that it is likely to cause permanent eye damage, even blindness – as many eighteenth-century navigators could have attested to, having to constantly sight the Sun in unfiltered sextants.

forces. It also implied that the Sun's interior must be some kind of dynamo.

The sunspots are not just cosmetic; they are the seats of solar activity. They are formed when the Sun's magnetism rises from within to bulge up into the solar atmosphere. Where they puncture the surface, they cool the gas from 6000 °C to about 4000 °C. The cooler gas emits only one-quarter the amount of light as its surroundings, so it appears dark by comparison.

Large groups of sunspots forewarn of gigantic solar flares that can unleash a billion times more energy than an atomic bomb. These happen when the magnetic fields producing the sunspots collapse. The colossal release of energy can trigger giant eruptions of solar gas that blast away a billion tonnes of the Sun's atmosphere to barrel through space, carrying electrical and magnetic fields. If these vast magnetic cannonballs strike the Earth, they spark aurorae and can damage electronics.

The largest storm of this type took place on 2 September 1859. It is known as the Carrington event, after the astronomer who witnessed the flare. Back then, the electronic technology of the day was the telegraph, which stopped working in spectacular fashion as the aurorae spread across the sky.

Phantom electricity from the solar storm began to surge through the telegraph wires. Operators were stunned unconscious, offices burst into flames. Simultaneously, compass needles spun uselessly under the influence of the aurorae. Global communications and navigation effectively ground to a halt and no one had the means to prevent it.*

* A detailed treatment of this apocalyptic event and its aftermath can be found in my book *The Sun Kings* (Princeton).

As the nineteenth century turned into the twentieth, Greenwich astronomer Walter Maunder and his wife, Annie,* began studying something dramatic that had happened to the sunspots in the latter half of the seventeenth century, at around about the time when modern Newtonian science was being born.

During this time it was unusual to see sunspots at all. Whole years went by without a single one being sighted. When they did put in an appearance, they were the rare exceptions rather than the rule. Indeed, throughout this grand minimum of activity, which lasted from 1645 to 1715, sunspots had been all but absent from the fiery surface. Clearly the Sun's energy generator continued to work, as it continued to shine, but the magnetic dynamo had faltered.

This grand minimum coincides with the worst years of the so-called Little Ice Age. This name itself is somewhat misleading because it was not a time of unremitting globally cold temperatures; instead, northern Europe was beset by a greater-than-average number of savagely cold winters. This is the time when the Thames regularly froze over and carnivals known as frost fairs were held on the icy concourse. Less benignly, the French army used similarly frozen rivers as thoroughfares to invade the Netherlands.

Although northern Europe was the most affected, other places felt it too. Iceland became locked in miles of sea ice and the island's population halved. New York harbour froze over,

* The pair met at the Royal Greenwich Observatory, where Annie was assigned to be Maunder's assistant. After Maunder's first wife died, the two began a relationship and eventually married. By the custom of the day, Annie was then obliged to leave her job. They continued to work together – Annie was a much better mathematician than Maunder – although frequently their joint work was only ever published under his name alone.

allowing people to walk from Manhattan to Staten Island.

Nor was this the first time that temperatures had plummeted. Between 1420 and 1570, a climate downturn struck against the Viking colonies on Greenland, turning them from fertile farmlands into arctic wastelands. Astronomers found that this, too, roughly coincided with a plunge in the number of sunspots. So could the Sun be to blame? Specifically, was there a causal link with the solar activity that is responsible for generating sunspots?

For a journalist or an academic, investigating the links between solar activity and Earth's climate is fraught with controversy. The subject has become highly politicized. The 'party line' is that man-made pollution is creating a greenhouse effect that is catastrophically warming the planet. Unless governments intervene and somehow force industry to stop burning fossil fuels and releasing carbon dioxide and methane into the atmosphere, we are heading for global disaster. To suggest that the Sun is in any way responsible for what's happening in Earth's atmosphere lays one open to being branded a 'climate sceptic'. Yet clearly the Sun is doing something and it is the role of science to investigate it.

Models lie at the heart of every prediction that is made about the future of Earth's climate, and those models are only as effective as the data and the theories programmed into them. If there is a missing interaction that takes place between the Sun and the Earth, then the models will not give trustworthy results. It may only be a small correction, but anything that better aligns the models with nature has to be a good thing.

Before the industrial revolution and man-made greenhouse gas emissions, the bare fact is that our planet's climate used to change. So, while the industrial pollution must now be having an effect, so, too, must those previous factors. They are unlikely to have gone away. Certainly, the Sun hasn't. The quantity and the nature of its energy stirs our atmosphere into weather patterns. Any change in the Sun's output, however subtle, must have an effect here on Earth.

Therefore, the question must be: how can we disentangle the Sun's effects from those of man-made gases and determine to what level the Sun is affecting us? If we understand this, then we remove a source of confusion from the data, which will allow us to quantify more precisely the man-made influence.

If this were a problem being investigated in the laboratory, the methodology would be obvious. The physicist would turn off one of the sources so that only the other was affecting the system. But how do you turn off the solar activity? Turns out it was simple. We just had to wait for the Sun to do it for us. In 2007, the sunspots disappeared.

The year before, the Sun had begun to settle into its usual minimum of activity. No one expected that we would see many sunspots for eighteen months or so and then the Sun would rouse itself for a new cycle. This would be cycle 24, as dated from 1755 when systematic records began.

Barely had the minimum set in when NASA-funded researchers made a sensational announcement: cycle 24 would be one of the most active solar cycles on record. According to their computer models, the Sun would be 30 to 50 per cent more active than the cycle that had just ended. That would place it on a par with the most active cycle on record, which occurred in the late 1950s to early 1960s at the dawn of the space age. And that meant a magnified threat of large solar flares.

Chillingly, our modern reliance on electronics for communications and navigation makes us more vulnerable than ever. We don't know when another Carrington-class flare might happen – perhaps once every couple of hundred years on average – but as technology progresses, electronic components become smaller and more sensitive, so they become more vulnerable to lesser solar flares.

That being the case, the NASA prediction of a 'doozy'* of a solar cycle was somewhat grim news (both technologically and linguistically). The Sun, however, had other ideas.

Little happened during 2007 and 2008. That's when eyebrows began to twitch because the dearth was more profound than anyone was expecting. If the coming cycle was going to be big, activity should ramp up quickly, but the Sun was spot-free 73 per cent of the time in 2008. This was extreme even for a solar minimum. You had to look back to 1913 to find a more pronounced minimum, when 85 per cent of that year was clear.

As 2009 arrived, solar physicists again hoped for some action. They didn't get it. At least, not until mid-December, when the largest group of sunspots to emerge for several years appeared. Was this, finally, a return to normal? Not by a long chalk. Cycle 24 has been whatever is the opposite of a doozy: a dozy, perhaps?

Significantly for our understanding of the Sun's effect on regional weather patterns, the Earth responded in an eerily familiar way. The winters of 2008–9 and particularly 2009–10 were unusually cold in northern Europe. Coming towards the end of a prolonged lack of sunspots, it struck some researchers

* http://science.nasa.gov/science-news/science-at-nasa/2006/10mar_stormwarning/

as too much of a coincidence with the Little Ice Age.

Mike Lockwood at the University of Reading, UK, led an investigation into average winter temperatures in the UK using the Central England Temperature dataset, which is a collection of meteorological observations that stretches back to 1659. He compared it to records of the highs and lows in solar activity, and found that, during years of low solar activity, winters in the UK were more likely to be colder than average. Significantly, when Lockwood removed the estimated climate warming due to industrial emissions from his models, the statistical link between solar lows and extreme winters grew stronger, suggesting the phenomenon is unrelated to man-made climate change.*

But to really prove a link, scientists would have to uncover a bona-fide chain of cause and effect that transmitted the changes in solar activity from the Sun across space to the Earth, and into the weather-bearing atmosphere. This would be difficult.

Spacecraft measurements show beyond doubt that the Sun is a remarkably constant emitter of energy. For more than thirty years, high-quality detectors have shown that the total energy output changes by only a tiny amount with the sunspot cycle: approximately 0.1 per cent between solar maximum and solar minimum. This translates into a difference of just 1.3 watts hitting each square metre of the Earth.

To place this in context, during solar minimum, about 1365 watts hits every square metre, while during maximum

* Lockwood, M; Harrison, R.G.; Solanki, S. K., and Woolings, T.: 'Are Cold Winters in Europe Associated with Low Solar Activity?' 2010: Environmental Research Letters 5doi:10.1088/1748-9326/5/2/024001.

that rises to about 1367 watts per square metre. So, in terms of total energy it's next door to nothing, but when it comes to the atmosphere not all solar radiation is created equal.

We used to hear a lot about the Earth's ozone layer. Ozone is a molecule made of three atoms of oxygen, as opposed to the two-atom molecular oxygen that keeps us alive. The ozone layer rests 20 kilometres above our heads – about twice the cruising altitude of a passenger jet – and is found at the boundary between the troposphere, the weather-bearing layers of the atmosphere, and the more rarefied stratosphere. Importantly for life, the ozone layer absorbs much of the harmful ultraviolet radiation that the Sun emits. The tiny fraction that does slip through is what gives us suntans and provokes skin cancer.

Since 2003 spaceborne instruments have been measuring the intensity of the Sun's output at various wavelengths and looking for correlations with solar activity. The indisputable fact is that ultraviolet light is strongly linked to solar activity. It helps carry away the explosive energy from solar flares and that means it varies much more strongly than we expected. In times of enhanced activity, the Sun can be a hundred times brighter at ultraviolet wavelengths than when there is little solar activity.

Although in terms of total energy this variation is all but drowned out by the constant mighty flux at visible wavelengths, the Sun's ultraviolet punches above its weight in the ozone layer. More ultraviolet light reaching the stratosphere means more ozone is formed. And more ozone leads to the stratosphere absorbing more ultraviolet light. So in times of heightened solar activity, the stratosphere heats up, driving faster winds up there.

The most famous stratospheric wind of them all is the jet stream. This is the shaft of air that blows from west to east across Europe. It can give transatlantic plane travellers a healthy tail wind when they are travelling eastwards. It also creates a 'fence' that separates the colder air to the north from the warmer air to the south.

Back in 1996, climate scientist Joanna Haigh, from Imperial College, London, showed that the temperature of the stratosphere influences the behaviour of the jet stream.* Lockwood's latest study makes it clear that when solar activity is low, the jet stream becomes liable to break up into giant meanders. These curve down over Europe, blocking the warmer westerly winds and preventing them from reaching the northern countries. Simultaneously, these teardrop-shaped twists open channels from the north, allowing Arctic winds from Siberia to howl down and dominate Europe's weather.

Hence solar activity does not specifically heat up or cool down the planet, but it distributes the weather differently around the globe. And Europe is particularly susceptible. Without factoring this out, it can bias our climate models because there are so many more weather stations in Europe. In short, we could be misinterpreting solar activity effects as global climate change. These could be either masking or artificially enhancing what we perceive as climate change. Either way, we need to know and climate models are now being changed to include these newly perceived effects.

At the same time as climate modellers are trying to understand the future of the Earth, solar modellers are making their own attempts at predicting the future of the Sun. In

* 'The Impact of Solar Variability on Climate', Joanna Haigh, *Science*, 17 May 1996, Vol 272, pp 981.

particular, is it heading for another Maunder Minimum?

On the face of it, the signs are not good. After one of the weakest solar maxima for a century, which took place in 2013–14, the Sun registered its first spot-free day for three years on 17 July 2014. There is often a brief revival of solar activity a year or so after solar maximum, but the next solar minimum cannot be too far away. Then what?

According to models, the strength of the solar cycle is governed by the circulation of two vast conveyor belts of gas that endlessly circulate material and magnetism through the Sun's interior and out across the surface. Known as the meridional flows, one works in each hemisphere, sweeping gas and magnetism away from the equatorial regions and up to the poles. Once there, the flow turns down into the Sun and sinks to a layer known as the tachocline. According to the models, the tachocline is a shell of hot gas 35,000 kilometres thick and buried some 200,000 kilometres deep inside the Sun. This layer is made of plasma – a gas so hot that electrons break away from the atom, leaving behind a positively charged 'ion'. Importantly, a plasma can carry or generate magnetic fields.

Beneath the tachocline, the Sun is so dense that it rotates as if it were a solid object. Above the tachocline, different latitudes rotate at different speeds. This behaviour continues all the way to the surface, where sunspots are swept around the Sun's equatorial regions in twenty-five days, whereas at the poles, it takes them a more leisurely twenty-eight days.

As the meridional flows journey towards the Sun's equator along the tachocline, the magnetic field is rejuvenated and produces new sunspots. This creates a feedback mechanism that can either strongly amplify or diminish the overall strength of the Sun's magnetic field. For the past fifty years the field has been building, and the Sun has been experiencing a period of unusually high magnetic activity.

On average it takes gas about forty years to complete a circulation, but adjusting the models to fit the latest data suggests that the surface portion of this flow has been slowing down, so the magnetic corpses of more recent sunspots have not been buried for magnetic resurrection so quickly. Hence, the solar cycle is withering before our eyes.

William Livingston at the National Solar Observatory in Tucson, Arizona, has been measuring the strength of sunspot magnetic fields for decades. In 2010, he and colleague Matt Penn sounded a warning by pointing out that the average strength of sunspot magnetic fields has been sliding dramatically since 1995. If the trend continued, they warned, the field will have slipped below the threshold needed for sunspots to form by about the beginning of the next cycle and we will be in Maunder Minimum territory.*

To everyone's relief, Livingston's latest measurements show that the field strength is no longer dropping but is hovering around the 2010 level. While this means that the chances of a Maunder Minimum appear small, there is also not much chance of levels returning to the highs of the previous few decades.

Even the most optimistic of models predict that cycle 25 will be one of the weakest on record. And that means northern Europe will be heading for a greater-than-average chance of more severe winters.

It is not just the immediate future of the Sun that astronomers have problems understanding; modelling the past is

* 'Long-term Evolution of Sunspot Magnetic Fields', Matthew Penn, William Livingston, arXiv:1009.0784v1 [astro-ph.SR].

pretty tricky too. For forty years they've been struggling with something called the faint young Sun paradox. It dates from the 1960s, when computing power allowed astrophysicists to run the first crude computer simulations of how changes in chemical composition affect the luminosity and heat output of Sun-like stars.

Eddington's work had established that the Sun shines by converting hydrogen into helium. A consequence of this is that there must have been significantly more hydrogen and less helium in the Sun's core when it was born 4.5 billion years ago. The models showed that this would have affected the early Sun's luminosity, lowering it by between 25 to 30 per cent of its modern value. That translates into an average surface temperature of the early Earth some twenty degrees cooler, or about ten degrees below the freezing point of water.

This is where the geologists got in on the act. They suggested dating how long it took for the Sun to be warm enough to melt Earth's water by investigating the oldest rocks on our planet. That's when the problems started, because even Earth's oldest rocks show it to be a far cry from the frozen wasteland suggested by the models.

For example, rocks from Jack Hills, Western Australia, are dated to 4.4 billion years ago and contain the mineral zircon. Inside the zircon are oxygen isotopes that point to it having being formed in the presence of liquid water. Then there are stromatolites, which are layered structures formed in shallow water by the action of microbial communities. Fossil stromatolites, also from Western Australia, date to 3.5 billion years ago, and constitute the best evidence yet for ancient bacteria.

Clearly, the early Earth was a clement haven for life.

The earliest proposal for a solution remains one of the

most popular: that some greenhouse gas allowed the early Earth to trap more of the weak Sun's rays and thus blossom. Astronomers Carl Sagan and George Mullen first made the suggestion in 1972. But finding the right combination of gases is proving tricky as many combinations contradict geological observations. Some think the solar model is just plain wrong. Others think that Earth absorbed different quantities of heat because of differences in the proportion of land to ocean to ice sheets way back when. Then there are the plainly frightening solutions.

In 2009, Jacques Laskar, an orbital dynamicist at the Paris Observatory, France, made headlines with a series of computer simulations that showed how the orbits of Mercury, Venus, Earth and Mars were not necessarily stable over billions of years.* In one particularly alarming scenario, Jupiter's gravitational influence could eventually fling Mercury outwards, risking collisions with Venus, Earth and Mars about 3.5 billion years in the future.

Inspired, David Minton, a planetary scientist at Purdue University in West Lafayette, Indiana,† investigated what it would take to have had Earth form closer to the Sun – just a few per cent closer would keep it warm enough – and only move out to its present orbit later.

He imagined a collision between two hypothetical planets that once belonged to our Solar System. According to his calculations this would have taken place about 2.5 billion years ago and the debris coalesced into present-day Venus. In the resulting gravitational flux, Earth was shoved outwards

* 'Existence of collisional trajectories of Mercury, Mars and Venus with the Earth', J. Laskar & M. Gastineau, *Nature* 459, 817-819 (11 June 2009) doi:10.1038/nature08096

† www.lpl.arizona.edu/~daminton/

to its present location, ensuring that as the Sun warmed, we didn't roast.

As Minton himself recognizes, testing the hypothesis is probably impossible. Although a planet's age can often be estimated by the density of craters seen on its surface (a bit like wrinkles on a face), Venus hides her age very well. A simple count of the craters on the planet suggests a figure of just 500 million to a billion years – far too young to be plausible under any scenario. This, planetary scientists are still working on how to extract a true age for Venus because they do not know what has happened to its surface to make it appear so young.

With so many solutions to the faint young Sun paradox, and scant evidence to help choose between them, scientists usually invoke something called 'Ockham's razor'. This is a little piece of philosophical wisdom attributed to the English Franciscan friar William of Ockham (1287–1347). Put simply, it means that among competing hypotheses, the one with the smallest number of assumptions is usually correct. In other words, applying this 'razor' allows you to cut away unnecessary intellectual complications.

Although named after the medieval friar, a version dates back to Ptolemy (AD 90–168), who stated that it was a good principle to explain phenomena with the simplest hypothesis possible. Others have tried to encapsulate its meaning in phrases such as 'Entities must not be multiplied beyond necessity' and 'It is vain to do with more what can be done with fewer.' These days, it is often stated as 'Keep it simple, stupid.'

It is a great lesson, and something that should be kept always in mind. The multiplication of assumptions and hypotheses is never in science's best interest because the resultant intellectual structures usually come tumbling down at some point.

As Newton wrote in the 1726 edition of his masterwork, *Principia*, 'We are to admit no more causes of natural things than such as are both true and sufficient to explain their appearances.'

As we shall discuss in later chapters, it is not always clear that cosmologists are strictly remembering this when they add new components such as dark matter and dark energy to their equations.

Whatever the precise details of the Sun and how we resolve the faint young Sun paradox, there is no doubt that by forensically piecing together their observations and the known laws of physics, astronomers have homed in on the knowledge that the Sun is a crucible in which the forces of nature go to war. It is a burning nuclear furnace where subatomic forces battle gravity and transform simple atoms into heavier ones.

All the atoms that make life possible – the carbon in our DNA, the iron in our blood, the oxygen in our lungs – all of it was built in the heart of massive stars that exploded billions of years ago, seeding space with these elements, ready to be incorporated into the stars that shine around us today.

And the Sun is just one of 200 billion stars in the Milky Way, the collection of stars that makes up our Galaxy. Gravity may be the architect of the Universe, but light is the fleet-footed messenger that makes the cosmos visible. It also carries secret information about the stars that astronomers began to unlock in the nineteenth century.

4
The Stellar Bestiary

Edward Charles Pickering did not suffer fools gladly. He was a bear of a man who ran the Harvard College Observatory in Cambridge, Massachusetts, during the latter decades of the nineteenth century and the opening ones of the twentieth. The observatory had been founded in 1839, and the first astronomer had been one William Cranch Bond, a Boston clockmaker. Although appointed to the job, he was awarded no salary; presumably the honour of the post was sufficient reward. As the years rolled by, the number of staff, telescopes and, thankfully, the worth of observatory salaries rose steadily.

By Pickering's time, the astronomers were being adequately recompensed and had embarked on a grand project: an entirely photographic survey of the sky, with each photographic plate recording hundreds of stars at a time. It placed Harvard at the epicentre of nineteenth-century astronomy.

The trouble was that this flood of data gave the astronomers the problem of keeping up with the analysis. It was laborious, painstaking work, and by all accounts the male assistants were simply not up to the task. Perhaps apocryphally, Pickering is said to have become so annoyed with them one day he exclaimed that even his maid could do better.

His maid was Williamina Fleming, a native of Dundee, Scotland, and a teacher by profession. She had emigrated to

America with her husband and their child but the marriage collapsed. To make ends meet, she was driven into domestic service. Fortuitously, she ended up with Pickering, who recognized her intelligence. So, at the age of twenty-four, she found herself working at the world-leading Harvard Observatory.

Initially hired for clerical work, one of her first tasks was to index the photographic plates. These were fragile objects, each one as big as a small tea tray and made of glass. They were unique objects too; one slip and the record of hundreds of stars could become nothing more than shattered fragments on the floor. Having devised an indexing system, she soon began classifying what was on them.

Each star had been turned into a rainbow streak by a prism placed in front of the telescope. Although the colours were lost on the photographic plate, the information they carried was clearly visible: a pattern of vertical dark lines superimposed on each spectrum. Had Fleming been born in modern times, she would have likened them to bar codes. Armed with a magnifying glass, she inspected each one and recorded the pattern of spectral lines.

Some stars displayed similar patterns, while others differed. It was as if the stars came from different families and could be grouped according to these spectral features.

The lines themselves had first been noticed in the Sun's light at the beginning of the century by English chemist William Wollaston, but he had dismissed their significance. Out of curiosity he had passed sunlight through a prism and magnified the spectrum. That was when he noticed that dark lines appeared to separate some of the colours. Because of the relative crudity of the telescope he was using, Wollaston saw only the four most prominent lines and he assumed they were natural gaps in the spectrum. With that thought, he ceased to pay them any more attention.

More than a decade later, the gifted telescope maker Joseph von Fraunhofer was more diligent. If Fleming's lucky break had been the break up of her marriage forcing her into domestic service, where she met Pickering, then Fraunhofer's lucky break had been the collapse of a Bavarian warehouse on top of him at the age of fourteen.

He was both an orphan and an overworked apprentice to a glass cutter when the accident happened. After hours of digging, rescuers unearthed the teenager and as recompense the Prince Elector of Bavaria, Maximilian IV, who had arrived to lend his support, began supplying him with books to help him learn his craft. He ordered the negligent glass cutter to give the young man time off to study. Within eight months, Fraunhofer had won a place at the Optical Institute of Benediktbeuren, a former monastery converted into a specialized glass-making facility.

By his mid-twenties, Fraunhofer had invented a machine for polishing lenses, as well as a new type of furnace that allowed him to make glass in larger quantities and to better quality than before. Indeed, it was a handmade Fraunhofer telescope that Heinrich Schwabe bought in the 1820s to begin his study of the sunspots (see Chapter 3).

In 1814, Fraunhofer invented the spectroscope, a single unit that combined a prism to split light into its constituent colours and a telescope to magnify the resulting spectrum automatically. One day, he rigged the spectroscope so that a shaft of sunlight entered the instrument. What he saw amazed him. As well as the familiar rainbow spectrum of colours, there were myriad dark lines along the spectrum. He began to investigate and soon counted 574 lines.

Next, he looked at the brightest stars and discovered that their light, too, was intercut with dark lines. While some stars displayed patterns resembling the Sun, others were different.

Chemists knew that different elements burnt with different-coloured flames: copper burns bluish-green, calcium red, and sodium with an intense yellow; these are as distinctive as any fingerprint. The technique of diagnosing a sample by burning a small proportion of it, became known as a flame test.

When passed through a prism, these colours resolved into bright lines, and people wondered what the link between these bright lines and Fraunhofer's dark lines might be. The man himself would never find out. On 7 June 1826, he succumbed to tuberculosis. He was thirty-nine.

Despite his single-handed advancement of telescopes, and the tantalizing presence of the Fraunhofer lines, some thought that the stars remained as unknowable as ever. In 1835, the French philosopher Auguste Comte wrote: 'We understand the possibility of determining their shapes, their distances, their sizes and their movements; whereas we would never know how to study by any means their chemical composition …'

It took until 1860 to prove Comte wrong.

At the University of Heidelberg, chemist Robert Bunsen and his assistant, Peter Desaga, had perfected a gas burner that nowadays bears the professor's name. It was perfect for flame tests because it burnt with an almost colourless blaze.

With the burner in hand, Bunsen now needed the best spectroscopes to analyse the light from the flame tests. That was where the physicist Gustav Kirchhoff came in. Tempted to Heidelberg by Bunsen, Kirchhoff set about perfecting the experimental apparatus while Bunsen used his chemical expertise to produce samples of unprecedented purity.

Together they investigated the spectral lines with confidence. Suspiciously, sodium light appeared to correspond with one of the darkest lines in the solar spectrum, a line that Fraunhofer had labelled 'D'. To tell them whether this was something deeper than mere coincidence, they devised an experiment.

Burning lime and passing the brilliant-white limelight* through a prism gave a continuous rainbow of colours with no dark lines anywhere. So Kirchhoff focused a beam of it through the flame of a Bunsen burner before it reached the prism. He then sprinkled sodium powder into the flame, making it flare brilliant yellow.

On the screen, he saw Fraunhofer's black D line appear in the spectrum of the limelight. He concluded that the sodium vapour had absorbed that specific yellow wavelength from the limelight and blazed it around the lab in the form of a yellow flame. The implication was staggering. The Fraunhofer lines were telling scientists what chemical elements were present on the Sun.

To double-check the idea, Kirchhoff performed a new experiment. Lithium burnt with a deep ruby glow, and there was no corresponding Fraunhofer line in the solar spectrum. So Kirchhoff directed a beam of sunlight through the burner and dusted the flame with lithium powder. There on the screen, nestled among the ubiquitous Fraunhofer lines, appeared a new one at exactly the red wavelength associated with lithium.

* The brilliance was so intense that lime was burnt in theatres to illuminate the stage. Hence the word 'limelight' has become synonymous with acting.

Breakthrough! Clearly, sodium was present on the Sun but lithium was not. By analysing these lines, the chemical composition of the celestial objects could be laid bare.

The publication of these results in 1860 spurred others to action. Early converts were the husband-and-wife team of William and Mary Huggins. They had built a backyard observatory in Tulse Hill, London, and immediately began using photography and prisms to capture spectral information about the celestial objects.

Physicists, too, were inspired and began charting the spectra of the known chemical elements. Bunsen and Kirchhoff took things further, discovering rubidium and caesium through their unique patterns of spectral lines. Other astronomers also played their part.

During the solar eclipse of 1868, Frenchman Jules Janssen and Englishman Norman Lockyer both spied a previously unknown spectral line. Lockyer proposed that it was coming from an as yet undiscovered element and named it helium, after the Sun god Helios, but it would be twenty-seven more years before chemists isolated the element on Earth, proving its existence beyond doubt.

By that time Williamina Fleming was at work classifying the stars by their spectral lines. Astronomers were regularly finding iron, calcium, magnesium, and many others elements in the stars. The photographic surveys were revealing the stellar Universe as a bestiary. There were stars of different colour, different size, and different brightness. Many stars were even changing their brightness in regular ways. To cope with this avalanche of data, Harvard Observatory director Pickering hired more and more ladies to assist Fleming. They were known as 'computers'.

In 1890, Fleming published her stellar classification catalogue. It was called the *Draper Catalogue of Stellar Spectra*, after Boston doctor and astronomy enthusiast Henry Draper, who made the research possible with a bequest. It contained 10,351 stars that Fleming had arranged into sixteen categories, labelled A–Q. The order depended upon the pattern of the spectral lines. For example, some stars showed very strong absorption lines due to hydrogen, and these were placed into categories A–D.

One of the Harvard computers was Annie Jump Cannon, a deaf physics teacher whose first love was astronomy. Even as a young woman, she had a competence about her and rose quickly through the ranks to oversee the spectral classification efforts. It soon became apparent that the original system needed reordering. Using the better data, she dropped most of the letter categories, and reordered the ones that remained. Now, the various spectral lines faded in and out across the whole scheme. She placed special emphasis on the lines of hydrogen gas, and by 1912, with her team classifying stellar spectra at a rate of 5000 per month, she had settled on the sequence O, B, A, F, G, K, M, which is still the accepted one today. She even suggested a mnemonic by which to remember it: 'Oh Be A Fine Guy/Girl Kiss Me!'.

We now know that this sequence mirrors the temperature of the stars. The O stars are the hottest with surfaces ablaze at more than 25,000 °C, whereas the M stars smoulder at just 2000 °C. The Sun is a G star with a surface temperature of 6000 °C.

Cannon wasn't the only new lady at Harvard to make her mark. There was also Henrietta Swan Leavitt. She, too, had been rendered deaf by illness, and her powers of concentration were admirable. Pickering set her to work cataloguing the variable stars. These were like beacons rising to maximum brightness, then fading again in a regular sequence that could last days, to weeks or months. As Leavitt discovered, and we shall discuss in Chapter 6, the variable stars have the most extraordinary application for gauging distances in the Universe.

By the early twentieth century, the Harvard computers had become the librarians of space, filing stars away according to their spectra and other observational phenomena. All of it relied on the transmission of information through starlight, and while the astronomers were happy to use light simply as a messenger, the physicists were interested in probing its very nature. In doing so, they got a lot more than they bargained for. They set the stage for the biggest coup in physics since Newton: Einstein and his theory of relativity. And it all started by asking how fast light travels.

For most of us, sight is our primary way of sensing our environment: we open our eyes, we see the world. We take light and vision for granted. Back in ancient Greece, the fifth-century BC philosopher Empedocles was deeply puzzled by the phenomenon and proposed that our eyes emitted the light we need to see the world. And he thought he knew why.

He proposed that everything was made of four elements: fire, air, water and earth. Different physical properties could then be achieved by mixing these fundamental substances in different combinations. In the case of an eye, he thought they were the concoction of the goddess Aphrodite, and that during her toil she mixed in a proportion of fire. This was the essential ingredient for vision: when we opened our eyes, light radiated into the world, which reflected off the objects around us and allowed us to see.

This idea persisted for the best part of a thousand years, despite a pretty hefty challenge from Euclid in the third century BC. He pointed out that we see the stars instantly when we open our eyes outside on a dark night and asked how that could be so, since the light from our eyes should have taken time to reach the stars and bounce back again.

No problem, said the proponents; clearly it was telling us that light travelled instantaneously from place to place, otherwise there would be a time lag from opening our eyes to seeing stars. And besides, haven't we all seen cats' eyes glinting in the dark?*

Fair enough, thought Euclid, and the status quo remained. This is a classic example of conservatism in science, where it is often seen as preferable to shore up an old idea with an embellishment, rather than scrap it in favour of a new idea. In principle, this is sensible because it allows the development of hypotheses and prevents the chaos of a constantly changing mindset. But the danger is that it leads to entrenchment, which is what happened in this case, especially when the great physician Galen bought into the emission theory in the second century BC.

A new explanation for vision began to dawn around the end of the first millennium AD. The year 965 saw the birth of Arabic scholar Ibn al-Haytham in Basra in modern-day Iraq. These days he is more widely referred to by the Latinized version of his name Alhazen, but as his work

* We now know that this is because cats' eyes are unusually reflective; they don't emit anything.

spread throughout medieval Europe he was usually referred to simply as 'The Physicist'.

His contribution was to experiment with light and prove that it only moved in straight lines. He used this information to investigate the eye as an optical instrument – a concept very far ahead of its time – six centuries before the invention of the telescope.

It seemed to Alhazen unnecessary to have the eyes emit anything when there were plenty of sources of illumination around already. So he set about detailing a theory of vision based upon his experimentation with light and his knowledge of the structure of the eye. But as good as his hypothesis was, it was no real use unless it offered a way of distinguishing it from the traditional emission theory.

Hypotheses that offer no predictions or tests are not useful if we are seeking certainty – a point we shall return to in Chapter 8 when we discuss a cosmological hypothesis called inflation – but for now, it is enough to say that to be true science, there must be a specific prediction that differentiates it from its rival.

For Alhazen's ideas to be taken seriously, the speed of light needed to be measured. The emission theory relied on light travelling from place to place instantaneously, so the nail in its coffin would be to prove that light travelled at a finite speed. The first person audacious enough to attempt the measurement was that pioneer of Italian telescopic astronomy, Galileo.

In 1634, Galileo was languishing under house arrest, having been forbidden to perform astronomy by the Roman Catholic Inquisition. He was all but broken, his supporters slowly nursed him back to health and then coaxed a final book from him, which was published in 1638. Known as the *Discorsi*, he summarized his life's work investigating motion. With a foot in both camps of philosophy and science, he believed that knowledge could be derived through reason alone and then backed up with experimentation and observation.

The book was written as a dialogue amongst three people, and in one of the conversations, Galileo describes a method for measuring the speed of light. He suggests having two people with covered lanterns standing on the brow of adjacent hills at night. One uncovers the first lamp, and the other uncovers theirs as soon as they see this. The first one times how long it takes to see the second lamp uncovered and this gives the time it takes for the light to travel to and fro. From this, the speed can be calculated.

Galileo had performed the experiment repeatedly with a separation of two to three miles but the answers were all over the place. We now know this was because a distance of a few miles is too close; the reaction times of the people conducting the experiment take far longer than the light's round trip. Several decades later, however, the Danish astronomer Ole Rømer was testing another of Galileo's ideas and inadvertently ended up deducing the speed of light.

He was observing the innermost of Jupiter's four moons, all discovered by Galileo in 1610. Known as Io, the moon completes a circle of its parent planet every 42.5 hours, and so frequently passes behind Jupiter and disappears from Earth's view. Galileo had suggested that by timing the precise moment of eclipse, an observer's longitude on Earth could be calculated. This would be because perspective effects would be important; one observer would be viewing Jupiter from a different angle from another, depending upon where each was standing on Earth to make the observation.

Working at the Royal Observatory in Paris, Rømer set to work. The trouble was that, as the observations mounted up, it was clear that the eclipses of Io were taking place later and later than expected; then the situation would reverse and Io would disappear into the shadow of its giant parent earlier and earlier. As he looked at the data, Rømer had an idea of why this might be happening.

He thought the movement of the planets was causing the variation in the eclipse time. As Earth revolved around the Sun during the year, it would draw closer and then further from Jupiter. These changes in distance would not affect the timing of the eclipse if light sped instantaneously from place to place. But if light travelled with finite speed, then as Earth drew further away from Jupiter, the light would take longer to cross the extra distance – just as a carriage would take longer to complete a journey between more distant cities – and so the eclipse would appear to happen later than scheduled. When Earth was moving closer, the opposite would be the case.

From these observations, Rømer concluded that light took about 22 minutes to cross the width of Earth's orbit, meaning that it would take 11 minutes for the Sun's light to reach our planet. That was not a bad estimate; the modern value pegs it at 8 minutes 11 seconds. It was the first proof that light travelled with a finite speed. As a bonus, it also gave the first observational proof that the Earth was in motion around the Sun, as Copernicus had suggested and Kepler had calculated.

Nevertheless, some people were still not convinced. It wasn't that they doubted the observations of the eclipses, but they seemed reluctant to believe the ramifications of light having a finite speed.

In England, Robert Hooke dismissed the findings by

saying that the calculated speed of light was so large that it was practically instantaneous. Not true at all. It stands at the root of some of the most profound observational effects in the Universe. The third Astronomer Royal, James Bradley, discovered the first of these in 1728, twenty-five years after Hooke's death.

As gentlemen of breeding tend to do, Bradley was spending the afternoon boating on the Thames. He noticed that the pennant attached to the mast changed direction when the boat turned, and wondered whether the same was true for starlight and the moving Earth. To understand why, imagine standing in the rain with an umbrella. With the rain falling straight down, the umbrella must be held directly aloft. Start moving, however, and the situation changes. Now, you are walking into the rain and so the umbrella has to be tilted forward to keep you dry. This happens because the rain falls at a certain speed.

Now think of starlight falling like rain at the speed of light onto the moving Earth. Telescopes will have to be tilted a little like the umbrella to allow the light to travel down the optical tube. Every six months, the Earth is moving in opposite directions, and so the direction of the telescope's tilt will have to be reversed. In practical terms, this means that the position of a star will appear to change every six months.

Bradley began a systematic search for the effect, now called aberration. He found it, measured it and used it to calculate that the speed of light was 10,210 times faster than the speed of the Earth through its orbit. From this he calculated that light must take 8 minutes 12 seconds to reach us from the Sun – just one second out from the modern value. But still not everybody was convinced.

While the astronomers are happy to draw conclusions just from observations, it makes physicists a little twitchy. They would much prefer to set up an experiment to make a direct measurement in which they can control everything about the set-up. So they started thinking again about Galileo's proposal to measure the round trip travel time. The trouble was that if Bradley were correct, the speed of light was so large that even a ten-mile round-trip would take just one ten-thousandth of a second – far faster than a human could react. So the project stalled until the middle of the nineteenth century and the birth of Armand Hippolyte Fizeau.

Born in Paris in 1819, Fizeau arrived in the world at roughly the same time as the invention of photography. Drawn to the newfangled discipline in his twenties, Fizeau soon started to develop a deeper interest in the behaviour of the light that photography relied upon, and he started to think about a way to measure the light's speed, rather than simply calculate it from observations.

The experiment he conceived was an update of Galileo's. Instead of humans, it relied on a spinning cogwheel that chopped the light into short pulses. When he got the speed exactly right, the pulse would travel to a distant mirror, bounce and arrive back at Fizeau at exactly the time that the wheel had advanced by one cog. Knowing the speed of the wheel and the distance to the mirror, he could calculate the speed of light.

Preliminary tests showed that, based on the speed he could turn the toothed wheel, the mirror would have to be at least five miles away. But where?

As luck would have it, his parents lived in Suresnes, a suburb of Paris almost six miles away from the centre. He decided his parents' place would host the mirror, and then lugged the rest of his apparatus up the hill of Montmartre in the north of Paris and began to set things up. This was decades before the construction began of the Basilique du Sacré-Coeur that dominates the summit today.

Fizeau waited for nightfall and then focused a beam of light through one of the gaps in the wheel and began to spin it. The wheel had a hundred teeth cut into its circumference and Fizeau rigged it to spin about a hundred times a second. Fine-tuning the speed, he watched until he could see the light returning from the mirror through the gaps in the cogwheel.

He found a value that was not as accurate as Bradley's but more in line with Rømer's, leading to others repeating and improving upon his experiment over the coming decades. Yet regardless of the details, it was now clear to both physicists and astronomers that light travelled at a finite speed and the speed that calculations finally converged upon was 186,282 miles per second.

Considering that the Earth's diameter is 7926 miles, it means that a ray of light could cross the equivalent distance of 23.5 planet Earths side by side in a single second. Clearly, this is an extraordinary speed but it is not infinite. The emission theory of vision was clearly wrong – and that wasn't the half of it.

The speed of light swiftly became a figure that any physicist would recognize instantly, especially if it turned up in an unexpected place, such as in a calculation about something thought to be completely unrelated. As coincidence would have it, this is exactly what happened for James Clerk Maxwell about a decade or so after Fizeau's experiment.

• • •

His school friends called him 'Daftie'. They were wrong. James Clerk Maxwell may have spoken with an accent that sounded rustic to the Edinburgh elite at his school. He may have been socially awkward on account of spending his first ten years on his father's Galloway country estate in the complete absence of children his own age, yet he was anything but daft.

He was innately curious. By the time he had learned to walk and talk, his curiosity had grown to insatiable proportions. He demanded to know exactly how everything worked. 'What's the go of it?' or 'Show me how it doos' was never far from his lips. Often unsatisfied by the answers he received from the adults around him, he developed the art of investigation. He would trace the bell wires from the servants' room through the hidden nooks and crannies of the house to the pull chords in the family chambers. Outside, he would do the same with the courses of streams. Doors, locks and keys fascinated him. Indeed, anything that moved or was mechanical captivated him completely.

At the age of thirteen, he astonished those who thought him daft by winning the Edinburgh Academy's mathematics medal *and* the school's first prize for English and poetry. A year later, he wrote a mathematical paper about how to draw complicated elliptical shapes easily, using pins, a pen and a length of twine. It was deemed worthy of being read to the Royal Society of Edinburgh, even though Maxwell himself was determined to be too young to attend and read his own work.

He was genius in the offing.

His ability emerged fully formed with a trio of scientific papers that he published through the Royal Society in 1865: 'A Dynamical Theory of the Electromagnetic Field' built upon the work of the great experimentalist Michael Faraday.

Until Faraday's experiments in the early nineteeth century, electricity and magnetism were thought to have been completely separate forces of nature. Before the twentieth century's recognition of nuclear forces, they stood alongside gravity and light as the means by which all phenomena in the Universe came about.

Through his masterfully constructed and executed laboratory investigations, Faraday showed that an electrical current generated magnetism, and conversely, a moving magnet induced an electrical current. Clearly, they were related in some way. Forces were the foundation of the Universe, according to Faraday. He saw no need for an ether in which light propagated. Instead, he believed some property of the force allowed it to transmit itself. But rather like Robert Hooke 150 years earlier, Faraday lacked the mathematical ability to transform his experimental measurements into an underlying mathematical description of the phenomena that would serve as a theory.

The person to do that was James Clerk Maxwell.

By 1865, he was thirty-four and at the height of his powers. His papers that year contained a set of equations, now called Maxwell's equations, that described the action of an 'electromagnetic field'. These are the equivalent of Newton's laws of motion but for interactions involving electromagnetism. Instead of pushes and pulls and gravity, Maxwell's work uses the attraction and repulsion of electrical 'charge' and magnetic 'polarity' to provide the force to make things move.

In the theory, electricity and magnetism are not different things but different aspects of the same underlying force. As part of this work, he calculated the speed at which a disturbance in an electromagnetic field would propagate. Astonishingly, it turned out to be the speed of light. Surely that's too much of a coincidence? he thought. It implied that not only are electricity and magnetism the same force, but that light is related as well. Each and every ray is a rippling wave motion of electromagnetic energy travelling at 186,282 miles per second.

In the same way that the gravitational attraction tells us about the distribution of masses in the Universe, so the light tells us about the distribution of electrical charge and magnetic polarity.

But hang on…

According to Maxwell, light speed was 186,282 miles per second, but measured against what? Everything else has a speed that is measured relative to something: a car's velocity is relative to the road; Earth's velocity is relative to the Sun. But light? What was that measured with reference to?

The obvious answer was the as yet undetected ether, the assumed medium through which light and gravity rippled. Indeed, Maxwell's calculations were taken as such a strong hint that the ether must exist that it triggered new interest in detecting this all-pervading stuff. But the trouble with the assumption of the ether is you can't detect something that doesn't exist, as a then ten-year-old Polish lad was destined to discover.

Albert Abraham Michelson had been born around the same time that Fizeau's speed of light experiments were taking place in Paris. His Jewish mother gave birth to him in Strzelno, Poland, where anti-Semitism was on the rise. When Albert was three, the family fled to the United States of America, settling in the gold-rush west of the country.

As he turned seventeen, his father showed him an announcement that the local congressman was supporting applications for students to attend the United States Naval Academy at Annapolis, Maryland. Albert jumped at the chance but was only admitted after President Ulysses S. Grant intervened. America's elected leader had been told that if he supported Albert's application it could help to turn California's Jewish population towards the Republican Party.

At the academy, Michelson excelled in physics. Upon graduation, he was assigned to the sloop-of-war USS *Monongahela*. It was during his two years at sea that he began to ponder a conundrum that Galileo had written about in his classic book *Dialogo*,* more than two centuries earlier. The great Italian physicist had discussed whether any experiment performed below decks in a moving ship could tell you how fast you were going. He concluded that there was not a single thing that you could do to measure the speed. If you were on a moving ship with no external clues, all you could determine was your own velocity relative to the ship.

Michelson was having similar thoughts but found little time to act on them. On his return to Annapolis, he was detailed to teach physics and chemistry at the naval academy. This involved recreating certain crucial experiments, including the measurement of the speed of light.

Instead of utilizing Fizeau's cogged wheel, he looked at a version developed by Léon Foucault, who had worked with Fizeau as a collaborator before becoming a rival. It used a rotating mirror. In the time it took for the light to make its

* This was the book that got him into trouble with the Roman Catholic Inquisition for asserting that the Earth moved around the Sun. He was tried in 1633 and found guilty of being 'vehemently suspected of heresy'. As a result, he spent the rest of his life under house arrest.

journey and return, the mirror rotated a little and bounced the returning light at a tiny angle. Measuring this angle allowed you to calculate the speed of light, so long as you knew the speed of the mirror's rotation. It was a big improvement in method because the mirror could be spun so fast and measured so accurately that Fizeau's five-mile-long arrangement was no longer needed. Foucault had performed it over just 60 feet.

As he began to set up his own version, Michelson saw places where he could make improvements, but to do that would cost more money than the Naval Academy was interested in paying. So he did the most sensible thing under the circumstances and went to see his father-in-law.

That same year, 1877, he had married the daughter of a wealthy New York stockbroker and lawyer. He persuaded his new father-in-law out of $2000, and invested it in the very best lenses, telescopes and mirrors that money could buy.

The resulting experiment took place on the banks of the river Severn in Maryland. The quality of the optics Michelson was using meant that he could push the baseline of the experiment from Foucault's 60 feet to 2000 feet. The increased distance, combined with the better equipment, resulted in a measurement twenty times more precise than Foucault's: 186,355 miles per second, to an accuracy of about 30 miles per second. The achievement was considered so impressive that it was reported in the *New York Times*, doubtlessly to the delight of his father-in-law.

Soon after his 1877 success, a stunning new thought hit him. There could be a way to measure your motion in a closed room after all. It relied on measuring the speed of light through the ether.

Although the ether had not yet been detected, scientists thought that it had to possess some pretty special properties.

First, it must be frictionless. As we saw in Chapter 1, Newton's laws of motion showed that the planets slid through their orbits without any resistance.

Second, it must be rarefied enough that sound could not propagate through it like the air. This was because we never hear sounds coming from space; everything observed there takes place in total silence. Seventeenth- and eighteenth-century experiments with air pumps had shown that sound could not propagate without air.

Third, and somewhat conflictingly, it had to be dense enough to convey light and the force of gravity.

Michelson's masterstroke was to realize that if the ether existed, the Earth's movement through space would create a headwind that should affect the speed at which light moved. Crucially, this was something he could measure. In his daughter's biography of him, titled *The Master of Light*, she recalled that he once gathered her and her brother together to explain the idea. Quite how much they understood at the time is unclear since they were both under ten when he conducted this now-famous experiment.

He told them to imagine a swimming competition between two perfectly matched components. One would swim across the river and back again. The other would swim up stream to the same distance as the width of the river, and then back downstream. If there were no current, then both would cover the same distance in the same time and the race would end in a draw. If there were a current in the water, then the swimmer going upstream would be impeded at first and then swept along on the return leg. The swimmer going across the river would have to angle his body slightly upstream so that he maintained a straight line across and back against the current.

A back-of-the-envelope calculation showed that the swimmer crossing the river would win because he was less affected by the current. So Michelson proposed to run this race with light, with one beam at right angles to the Earth's motion along its orbit and the other directly in line. When he did this in 1887 with his colleague, Edward Morley, he entered the scientific history books. There was no winner out of the two beams of light. Both returned at exactly the same time as one another. The ether simply did not exist, but so deeply ingrained was everyone's belief in it that even Michelson himself was reluctant to conclude this.

In fact, the conundrum was worse than that, and struck at the heart of science. Velocities are usually added together. This is why when two cars hit head-on, the collision speed is their combined velocities. Michelson and Morley showed that this was not true for light. The beams travelled at exactly the same velocity even though they were travelling in different directions relative to Earth's motion through space.

Light is the only thing in the Universe that behaves in this absolute way. No matter how you are moving, if you measure the speed of light it always comes out the same. Its velocity is hard-wired into nature. This is why Maxwell's calculations had just given a velocity without reference; it is the velocity of light that is relative to everything, no matter what other movement is involved. To this day we don't understand why this should be true but it has some amazing consequences. It changes our entire perspective of what the Universe is and how it behaves, and leads us into weird realms that defy the solidity of the world we take for granted.

And the person who opened our eyes to this new way of thinking was the man who now characterizes science: the twentieth-century icon that is Albert Einstein.

5

Holes in the Universe

Wild hair, bushy moustache, careworn lines around deep, dark eyes, tongue sticking out: this is the Albert Einstein that most of us picture. The famous tongue photo was taken on his seventy-second birthday in 1951 by a news agency photographer. Tired of smiling for the camera, Einstein made the face, photographer Arthur Sasse pressed the shutter, and history was made. A copy signed by Einstein was auctioned in 2009 for $74,324, making it the most expensive image of the scientist ever sold.*

However, the old, familiar Einstein is not the one who made the big breakthroughs. To find that incarnation, we have to rewind forty-six birthdays to 1905. In those days, the dark eyes were the same, but the frame was slim and the hair dark and well cut. He was certainly not the eccentric figure that later came to symbolize science, although throughout his life he never much liked wearing socks.

Back in 1905, he was married with a one-year-old son, and he was working in the Patent Office in Bern, Switzerland. In his spare time, he was completing his thesis under the

* Part of the reason for its value is Einstein's inscription, which states that the image was a gesture to all of humanity, and this has been interpreted as his comment on the rise of McCarthyism in America. During this time, there was increasing pressure on everyone, including academics, to inform on colleagues who were thought to have communist leanings. For Einstein this must have had unpleasant overtones of the rise of Nazism that drove him to leave Germany for America in 1933.

aegis of a physics professor at the University of Zurich. That should have been enough for anyone, yet in addition to being husband, father, patent clerk and PhD finalist, he published four extraordinary papers in the prestigious journal *Annalen der Physik* (*Annals of Physics*). As a result, this is now referred to as Einstein's *annus mirabilis*.

We have already mentioned one of these papers and the effect it had on astronomy: it was the paper that led to $E = mc^2$. Two others were theoretical explanations for puzzling experimental results and later became foundation stones for atomic theory and quantum mechanics. The fourth was entitled 'On the Electrodynamics of Moving Bodies'. In it, Einstein presents a version of his world-famous theory of relativity.

The word 'relativity' is now so closely associated with Einstein, and so redolent of highbrow difficulty that it is tempting to switch off when hearing the word. Yet, all relativity really means is that movement must be measured relative to something else. Recall Galileo's book *Dialogo* from the previous chapter that asserted no experiment can be performed inside a room isolated from its surroundings that will allow you to measure whether you are moving or not. This was an early recognition of relativity.

Five decades later, Newton swept it aside. He framed his laws of motion and gravity in the belief that there was an absolute framework of space and time against which all motion could be measured. However, the results of the Michelson–Morley experiment showed that Newton was mistaken. Einstein took this as his inspiration and was convinced that measurements could only be made relative to something else; your state of motion made no difference to the outcome of any experiment that could be performed. In other words, the laws of physics remained the same regardless of your state of motion. This was why no isolated experiment could distinguish how fast you were moving.

Einstein believed this so strongly that he called it the principle of relativity, a principle being an idea that is used as a foundation stone for a theory.

He also took on board that the speed of light was invariant regardless of yours or anybody else's state of motion. In calculating the consequences of all this, he found that he predicted a set of weird phenomena that could be tested by experiment.

First, time and mass dilate or increase the faster we move. By dilate, I mean that time slows down; each second dilates to become longer and longer the faster our velocity becomes. This strikes the first blow against there being an absolute time, as Newton conceived it, ticking rigidly like some schoolmaster's pocket watch, while the drama of the cosmic playground unfolds.

Second, length contracts, so objects appear narrower the faster we pass them by. This strikes at the Newtonian concept of space as rigid and fixed.

Indeed, Einstein showed that mathematically time and space were pretty much indistinguishable. He spliced them together in his equations and referred to them as 'spacetime'. He was not the first to feel this intuitively. It had been familiar to philosophers, scientists and poets for a long time before.

H. G. Wells had referred to the concept in his 1895 science fiction novel *The Time Machine*, stating that any real object must extend in four dimensions: length, width, height and duration. Edgar Allen Poe had claimed that space and time were the same in his 1848 non-fiction prose poem, *Eureka*.

What Einstein did was fashion the concept into mathematical form. He called it the special theory of relativity and illustrated one striking example of its consequences with the use of a train. Imagine Alice, her husband Bob, a train and a spectacular act of God. The situation is this: Alice is on the train and Bob is standing on the embankment watching it go by. All of a sudden, two vivid bolts of lightning crack the sky. Simultaneously, one hits the front of the train, the other hits the rear. The train rumbles on, and Bob shoots off to the station to pick up Alice (who is presumably somewhat grumpy that he didn't make his way straight there instead of standing like an idiot on the embankment waving).

Truculence disposed of, they talk about the lightning and Bob voices his incredulity that the two bolts struck at exactly the same time. Alice is now convinced more than ever that Bob is an idiot, and while contemplating making some serious life choices relating to him, puts her point of view. According to what she saw, the lightning struck the front of the train first, then the rear. Surely they either happen at the same time or they don't. Who is right? Einstein showed they are both correct. The key to understanding is that there is a relative speed between them.

On the embankment, Bob was stationary with respect to the Earth and the light rays from the lightning strikes happened to reach his eyes at the same time. Hence, he saw them hit the train simultaneously.

On the train, Alice was being swept forwards by the velocity of the train relative to the Earth, and she was swept into the light from the forward strike slightly before the light from the rear strike had time to catch up. So, from her perspective, the lightning strikes were not simultaneous.

The only way for them to agree that the two events they saw were the same is to use a formula that Einstein derived from his theory, but which is known as the Lorentz transformation.* This is the centrepiece of special relativity and takes full account of observers who are moving with different velocities. In doing so, it utterly destroys the concept of *absolute* time and space because reality is what the observers see and measure, and that depends on their *relative* motion.

Einstein's choice of trains in his thought experiment is probably no coincidence. In 2000, Peter Galison, a science historian at Harvard, published a paper† in which he argued that Einstein's initial ideas that led to relativity may have been helped or even wholly sparked by his work at the patent office. Railways had become the arteries of Europe and that meant people needed to know what time the trains were going to arrive and depart.

Up until this point, clocks had been set locally either by reference to the Sun or to other clocks. A few minutes deviation from place to place was of absolutely no consequence. But trains coming in and out of stations changed all that. If a train was due to arrive at 7 o'clock, then stations up and down the line needed to synchronize their clocks.

Einstein worked on a lot of patent applications that tried to solve this problem using mechanical devices to tie clocks together using electrical pulses carried along wires between

* Hendrik Lorentz (1853–1928) derived the transformation equation to fit the results of the Michelson-Morley experiment. The transformation suggested that bodies in motion were compressed, though why this should happen was a mystery. Einstein's discovery of the Lorentz transformation as a natural consequence of special relativity provided the theoretical underpinning.

† 'Einstein's Clocks: The Place of Time', Peter Galison, *Critical Inquiry*, Vol. 26, No. 2 (Winter, 2000), pp. 355–389.

stations. He inevitably began to wonder about the problem as well. Typically for a scientist, he stripped it down to the basics and found himself with a simple question: What do I mean when I say the train will arrive at 7 o'clock?

In solving that, he derived special relativity. In other words, the greatest recent breakthrough in our understanding of the Universe came about, not through an existential inspiration to know the cosmos, but from a simple desire to keep the trains moving on time. And the great man was not done yet. It wasn't special relativity that made Einstein famous, it was what came afterwards: the general theory of relativity.

Special relativity is so-called because it only applies in certain special circumstances. Those special circumstances are when things are moving with set velocities or are stationary. The equations would not work for an accelerating object, so Einstein knew that there was a deeper theory to be had, a more general theory that applied to all forms of motion. And here a grand prize awaited him. Einstein knew that a general theory of relativity would also be a theory of gravity because it was impossible to tell the difference between accelerated motion and the action of a gravitational field. Think of it like this. Imagine that you are in a lift, which is essentially a closed room with no window to the outside world. There are four different circumstances to consider.

In the first, the lift is suspended in a lift shaft and gravity is holding you to the floor.

In the second, the cord is cut and the lift goes plummeting down the shaft with you inside it. Because you are now in a state of free fall, you feel weightless and float around inside the lift.

In the third case, the lift is taken into space, well away from any celestial body generating gravity. You float about because there is no gravity to pull on you.

In the fourth case, a rocket motor is strapped to the lift and turned on to accelerate you at a comfortable level. Inside the lift, you stop floating and feel heavy again. You can stand on the floor.

From inside the lift, there is no experiment that you can do to discriminate between cases one and four. Similarly, cases two and three are indistinguishable from inside. It is known as the principle of equivalence. There are several different versions of it, but they all boil down to one idea: that the effects of gravitational fields are indistinguishable from the effects of accelerated motion.

The principle of equivalence underpins all gravitational motion throughout the Universe, helping to dictate the speed at which planets orbit their stars and stars orbit their galaxies, and Einstein balanced his gravitational theory upon it.

But why would Einstein want to do this when Newton had supplied a perfectly good theory of gravity? The answer was that astronomers had found a crack in Newton's work. After a few centuries of triumph, one observation was beyond its ability to explain. And in science, only one observation is needed to kill a theory, or at least show that it is incomplete.

The fly in Newton's ointment was the planet Mercury.

Throughout the nineteenth century, telescopes became more and more precise and astronomers found themselves able to measure the night sky with unheard of accuracy. That's when the problem with Mercury's orbit became apparent. The planet was not moving as Newton's laws prescribed. Every time it dipped close to the Sun, the direction of its orbit was slightly changed. The effect is known as the precession of perihelion, and Newton's theory gave the wrong figure for this.

Astronomers had spent decades searching for an undiscovered planet closer to the Sun that they thought might be pulling Mercury off course. Yet try as they might, it wasn't there. While some tried to tinker with solutions involving dust dragging on the planet, Einstein became convinced it was a problem with our understanding of gravity.

He hoped that his new theory would show that, close to the Sun, gravity behaved slightly differently from the way Newton predicted, but it was gruelling work. The mathematics was beyond his immediate abilities and the years passed until it was 1914, almost a decade after his *annus mirabilis*. Einstein was no longer at the Patent Office in Bern, and there had been many other changes in his life, not all of them good. Following his triumph with special relativity, he had moved to the University of Zurich, but in 1914 he had been tempted to the University of Berlin with the promise of his own research institute, staff, and light teaching duties. He also had a secret reason for going.

His marriage was crumbling, and he had fallen in love with his first cousin, a widow who lived in Berlin with her two daughters. Although Einstein took his wife Mileva and their two boys with him, he soon packed them off back to Switzerland. Then the war came. Amid the overwhelming nationalism, Einstein alienated himself from his colleagues by opposing the German aggression. Cold-shouldered by many around him, Einstein found himself with little else to do except to retreat to his garret and work on his hypothesis. Whenever the mathematics ran aground, he backed up, changed the formulation slightly, and tried again.

In Newton's theory, gravity is like the tension in a piece of string between two objects. As they move around one another, the tension and therefore the pull of gravity remains the same. Einstein found that this wasn't quite true; in his own formulation the tension could vary from place to place. Instead of a multitude of strings of different tension emanating from the central object, it was better to think of it as a landscape, with hills and valleys. This was the spacetime continuum. The deeper the valley, which was termed a gravitational well, the stronger the gravitational field. The gradient of the valley's slope could change from place to place, indicating the speed with which the force of gravity was changing.

In November 1915, Einstein succeeded. Overlooked by portraits of Newton, Maxwell and Faraday, he calculated what the contours of the spacetime continuum should look like around the Sun, near Mercury, and used it to derive the precession of the innermost planet's perihelion. It matched the observed value perfectly. Einstein nearly died of shock, suffering heart palpitations in his excitement.

He had stepped beyond Newton, showing that the Englishman's theory was valid only when the gradient of spacetime was shallow and therefore the strength of gravity was relatively weak. But in stronger gravitational fields, only Einstein's theory predicted the corrections that needed to be made.

Nevertheless, to be taken seriously, it wasn't good enough simply to explain what had already been seen. By the strict definition of the difference between a hypothesis and a theory, he needed to use the mathematics to predict a phenomenon that hadn't been seen or measured yet. Einstein had one, but only through a piece of good fortune.

In the Sun's gravitational well, starlight would stop following straight lines. It would instead follow the contours of spacetime and this would mean it would change course, like a golf ball lipping the hole. The phenomenon became known as gravitational lensing but was almost unobservable because the Sun is so bright that stars whose light passed close enough to be affected would be lost in the glare; it was like trying to see a firefly next to a searchlight. The only chance was at a total eclipse, when stars could be seen close to the Sun because the Moon blocked out its light.

Shortly after arriving in Berlin, Einstein had arranged for an eager young astronomer called Erwin Freundlich to travel to the total eclipse visible from the Crimea to look for this deflection, giving him another number to aim for in his calculations. Disastrously for Freundlich, the First World War broke out just after his arrival, making him an enemy citizen in Russian territory. He was taken prisoner, his telescope and cameras were seized, and it took months to secure his safe return to Germany. The equipment remained impounded for the rest of the war.

For Einstein, this was the best thing that could have happened. If Freundlich had been successful and returned with the measurement, Einstein would have used it in the theory's formulation. Now, however, he could predict its value and it could be tested. It was general relativity's crucial test.

Cambridge astrophysicist Arthur Eddington enters our story again at this point. He wanted to test Einstein's ideas and that meant travelling to the total eclipse that was coming up in 1919. Together with Astronomer Royal Frank Dyson, Eddington persuaded the British government to back them. They argued that since this work had the power to overturn the 'English' theory of Newton, it was better to have some national claim on the new idea as well. They were also careful to hide the fact that it was the work of an essentially German (therefore 'enemy') physicist.

Suitably convinced, the government excused Eddington from war work in order to make preparations. As it happened, the war was over when Eddington set sail in 1919 for the African island of Principe. There, the weather was less than promising, with heavy rain on the morning of the eclipse. At the crucial moment, however, the clouds parted and Eddington took his images.

That night he developed and studied them. Some were blurry because of the clouds. He discarded these and concentrated only on the good ones. The deflection he measured was entirely consistent with what Einstein had predicted. As impossible as this strange invisible landscape of the spacetime continuum sounded, Eddington's work proved that it had to be true in some way.

Announcing the discovery in London brought the press flocking from all over the world. *The New York Times* lacked a science correspondent at the time but sent what I presume they thought was the nearest thing: their sports correspondent. In keeping with many at the time and afterwards, he didn't really understand general relativity, but what he lacked in understanding, he made up for with enthusiasm and he wrote one of the greatest sets of headlines in journalistic history:

- **Lights All Askew in the Heavens**
- **Men of Science More or Less Agog Over Results of Eclipse Observations**

- **Einstein Theory Triumphs**
- **Stars Not Where They Seemed or Were Calculated to be, but Nobody Need Worry**
- **A Book for 12 Wise Men**
- **No More in All the World Could Comprehend It, Said Einstein When His Daring Publishers Accepted It**

And the legend of Einstein was born. The world was heartily sick of the old absolutes. The imperial powers and the status quo had led to the carnage of the First World War. People wanted change and here was a scientist who understood the Universe at a deeper level than anyone before him: and he said that everything was relative. No more absolutes. It was a progressive way of thinking: a new philosophy for a new age, both for science and society.

For science, general relativity opened up the study of the Universe as never before. Einstein's field equations, a set of sixteen mathematical expressions that described the space-time landscape, could be applied to any scale. They applied equally well in the vicinity of an individual celestial object as to the Universe as a whole, and that meant, for the first time in history, humanity had an equation that could describe the whole Universe and show how it changed with time. Using it, they could reconstruct the history of the cosmos, and investigate its future. It signalled the true beginning of cosmology. But before this particular discipline took form, there was another strange prediction from general relativity to deal with, one that Einstein didn't see coming and wasn't initially prepared to believe.

• • •

Back in 1916, while the war had still been raging, a German artillery officer had found an utterly remarkable possibility in the swirl of figures and symbols. The man was Karl Schwarzschild, and what he had found theoretically, we now call black holes.

Schwarzschild himself never needed to go to war. At forty, he was considered an old-timer but volunteered anyway. He served on both western and eastern fronts and swiftly achieved the rank of lieutenant. Behind him was a solid corpus of scientific work; ahead lay his greatest achievement and an untimely death.

Einstein's paper announcing general relativity to the world was published in November 1915. By this time, Schwarzschild was on the Russian front and suffering from a rare skin complaint called pemphigus. This disease manifests itself as a series of blisters that turn into painful sores. To take his mind off his condition, and the heavy fire experienced in his sector of the front, he read the physics journals.

General relativity astounded him. In a letter to Einstein, written three days before Christmas that year, he called the theory 'an entirely wonderful thing'. The letter contained more than just praise.

Einstein had published only an approximate solution to his sixteen equations. Indeed, the great man thought that an exact solution was probably impossible because of the apparent complexity of the mathematics. Schwarzschild found that there was a straightforward way to solve the equations and find exact numerical descriptions for the contours of the spacetime continuum around stars or other celestial objects. The only approximation he had to make was that the celestial object was uniform and non-rotating.

Einstein had come unstuck by using a rectangular system of *x*, *y* and *z* coordinates. Schwarzschild replaced this with a polar coordinate system that relied on two angles and a distance, but there was a sting in the tail of his triumph. As well as calculating the geometry of spacetime for a large symmetrical celestial object, he also performed the calculation with all the mass squeezed into a minuscule point. When he did so, something amazing happened.

The equations became impossible to solve at a certain radius. As one neared the central object, some of the mathematical terms became infinite. This radius became known as the Schwarzschild radius and immediately posed the problem of what was its physical meaning.

Schwarzschild did not live to take much part in the search for an answer. He died on 11 May 1916, aged just forty-two, probably of the skin condition that plagued the last year of his life. Had he lived, he would have been an old man by the time the solution did finally come to light. It was presented forty-two years later by David Finkelstein, a physics professor at the Massachusetts Institute of Technology.

Finkelstein recognized that mathematically the Schwarzschild radius looked like a valve in spacetime, across which things could pass only one way. Once they fell in, the gravitational pull inside the Schwarzschild radius was so strong that nothing could escape. Indeed, the gravitational field was so strong that nothing could withstand it. The inescapable conclusion was that anything crossing the Schwarzschild radius would be crushed out of existence. But what should such an exotic celestial object be called? In the scientific literature of the time it was referred to as a 'gravitationally completely collapsed object'.

Factually accurate but hideously unglamorous, this mouthful had to change. No one quite knows who coined the term 'black hole', but it was definitely being used quite widely at the January 1964 meeting of the American Association for the Advancement of Science* because journalists for *Science News Letters*† and *Life*‡ magazines both used it in their coverage.

As for where the term came from in the first place, look no further than Imperial India of the eighteenth century. The 'Black Hole of Calcutta' was a phrase used to describe an infamous dungeon in Fort William, Calcutta. The inescapable nature of the gravitational prisons that astronomers were contemplating may have led them to appropriate the name. It was also in the eighteenth century that science first flirted with the idea of a black hole. Back then they were referred to as dark stars.

The discussion originated with an English clergyman and natural philosopher, John Michell. Working in 1783, a century and a half before Einstein's general relativity, he based everything on Newton's theory of gravity and the measured speed of light. He reasoned that light would have to work against the pull of gravity to escape its parent star. Since the escape velocity would be higher for larger stars, he suggested that there could be stars large enough that the escape velocity exceeded that of light. Hence, the star would be producing light that could not escape into space.

* See *Black Hole*, Marcia Bartusiak, Yale University Press (28 April 2015) for more information.

† *Science News Letters*, Vol. 85 #3, 18 January 1964 (https://www.sciencenews.org/archive/black-holes-space?mode=magazine&context=1668&tgt=nr)

‡ 'What are quasi-stellars? Heavens' New Enigma', *Life*, 24 January, 1964, p. 11.

In his paper to the Royal Society on the subject,* Michell suggested that, although the dark star could not be seen in isolation, there were plenty of binary stars in orbit around one another that were known. A dark star could be present in such a system, in which case the visible star would move in an orbit around the invisible companion, betraying its presence. When astronomers in the 1960s and 1970s were contemplating whether black holes could indeed exist, they returned to this idea.

It was 1964 and pioneering astronomers were attaching Geiger counters to the nose cones of a pair of rockets. These so-called 'sounding rockets' were suborbital and just touched space before falling back to Earth. They were looking for X-rays that are blocked from reaching the ground by the particles in the upper reaches of the Earth's atmosphere.

Each flight lasted only a few minutes and the swathes of sky they covered were but thin strips, yet it was enough. Eight sources of X-rays were found in the sky, and one of the strongest was located in the constellation of Cygnus.

The astronomers called it Cygnus X-1 and identified it as coming from a star known only by its classification number HDE 226868. Although the star is a giant with a surface temperature of several tens of thousands of degrees, it is not hot enough to be emitting the deluge of X-rays that astronomers were seeing. The X-rays required something that could heat matter to millions of degrees.

With the mystery deepening, repeated studies of Cygnus X-1 and the other celestial X-ray sources became essential. In

* John Michell, *Philosophical Transactions of the Royal Society of London*, 1783, Vol. 74, p. 35. (see http://www.amnh.org/education/resources/rfl/web/essaybooks/cosmic/cs_michell.html for context)

December 1970, NASA's Christmas gift to the astronomical community was to launch Uhuru, the first X-ray satellite. The name comes from the Swahili word meaning 'freedom' and was used because the satellite blasted off from a converted oil platform sitting in the Indian Ocean, just off the coast of Malindi, Kenya.

Uhuru was a total success. It discovered more than 300 previously unknown X-ray sources in the Universe and allowed some long-term monitoring of Cygnus X-1. This showed that whatever was creating the X-rays was also varying its X-ray intensity several times every second. It all pointed to something smaller than our Sun yet much hotter and capable of rapid variation.

The clincher came a couple of years later when two groups of astronomers worked independently to apply Michell's idea about looking for orbital motion in the visible star created by the suspected dark star.

They split the light from HDE 226868 using a prism and identified its spectral lines. They watched as these lines moved first one way, to longer wavelengths, and then the other, to shorter wavelengths. According to the work of nineteenth-century physicist Christian Doppler, this was the signature of motion. The spectral lines would be squashed towards shorter wavelengths when the star was approaching an observer, and stretched to longer wavelengths when it was moving away. It was called the Doppler effect.

The fact that the star appeared to be both approaching and then receding in a regular repetitive way could only mean that it was in orbit around something. And in terms of gravity that something was enormously powerful. The giant star was pulled around its orbit once every 5.6 days. Even Mercury, the fastest-orbiting planet in our Solar System, takes eighty-eight days to complete a circuit of the Sun.

This period was reflected in the time it took for the spectral lines to shift one way then the other. It could be used with the amplitude of the spectral lines' movement, employing Kepler's laws of planetary motion, to give the mass of the dark star generating the gravity. It was huge, at least ten times the mass of the Sun. Yet it was invisible at optical wavelengths. After some discussion of possible sources of error, it was generally accepted by the astronomical community that Cygnus X-1 must be a black hole.

The cataract of X-rays was coming from superheated matter that was circling in a disc around the black hole, awaiting its final plunge into oblivion. The matter itself was gas that was being siphoned from the visible star to feed the black hole's monstrous gravitational craving.

It was impossible to observe the black hole directly – how do you observe something that is just a few kilometres across and absorbs absolutely every single ray of light that falls on it? Nevertheless, the weight of circumstantial evidence was so large that, by the mid-1970s, most astronomers were convinced that Cygnus X-1 was a black hole. One notable holdout was Stephen Hawking. It wasn't that he objected to Cygnus X-1 particularly, but black holes in general. His scepticism manifested itself in a bet made in 1975 with American theoretician Kip Thorne that Cygnus X-1 was not a black hole.

Hawking's reluctance to concede that Cygnus X-1 was indeed a black hole may have been because he knew very well that if black holes existed, it meant that there were places in the Universe that general relativity could not go. This is because at the very centre of the black hole is the singularity, the final resting place of all the matter that has fallen in. It is where the matter will be crushed out of existence – whatever that means. And that's the problem: what does it mean to have a point of infinite density but zero volume? General relativity cannot tell us; the mathematics becomes insoluble.

Perhaps it was better to hope that some as yet unknown physical phenomenon halted the collapse and stopped these holes appearing in the Universe. But nothing we've found in theory or reality appears to be able to prevent black holes forming.

By 1998, Hawking was ready to concede the bet. He thought that although the specific evidence for Cygnus X-1 had not changed, there was now so much evidence for other black holes dotted around the Galaxy and the wider Universe that there was no more room for reasonable doubt. Besides, as we shall see in our final chapter, he had bigger fish to fry and a new bet to concentrate on. So Hawking paid for Thorne's prize: a year's subscription to *Penthouse*.

Black holes such as Cygnus X-1 are known as stellar black holes; they are the most prevalent kind in the Universe. Containing several times the mass of the Sun, a stellar black hole is formed when a massive star explodes as a supernova at the end of its life.

Supernovae begin with the cessation of nuclear fusion in the heart of the star. The star's inert core, which models suggest will be about the size of our planet but contain as much mass as almost one and a half times the Sun, collapses in a split second under the weight of its own gravity to become an incredibly dense ball of matter just 10–20 kilometres across. Under such intense pressure, the electrons that provide the outer vanguard of the atoms will be forced inside the central nuclei. They will merge with the protons to create particles known as neutrons.

The rest of the star comes crashing down on top of this so-called neutron star, striking its surface and creating a shock wave that rushes outwards, sparking the supernova explosion. Behind the shockwave, the additional weight of matter on the core can increase its gravitational field so much that it collapses again to become a black hole with a Schwarzschild radius just a few kilometres across.

At the other end of the scale are the supermassive black holes, each containing something between a million and many billions of times the mass of the Sun. It is thought that a supermassive black hole sits at the heart of every galaxy and provides the gravitating fulcrum around which the massive collections of stars rotate.

Even though supermassive black holes contain such a vast amount of mass – they take up no more volume than an average solar system, often a lot less. Our own Galaxy's central supermassive black hole is known as Sagittarius A* (pronounced 'A-star'). Based on the movement of stars and gas clouds in its vicinity, it is estimated to contain about 4.5 million solar masses. This is all squeezed into an event horizon (the modern name for the spherical volume with the Schwarzschild radius) some 27 million kilometres across, which is about half the distance of Mercury from the Sun. From our vantage point on Earth, about 30,000 light years away, its silhouette would appear no larger than a football on the surface of the Moon.

In the Milky Way, as in about 90 per cent of all galaxies, the supermassive black hole is almost inert because nothing is falling into it. In the other 10 per cent, it is constantly feeding from surrounding celestial objects, and each of these drives an extraordinary engine of activity that can be seen across billions of light years of space.

The active galaxies release prodigious quantities of radiation and particles into space. It is like a hugely scaled-up version of the radiation coming from Cygnus X-1. Before matter plunges to oblivion, it circulates in an accretion disc, heating up. On the scale of a supermassive black hole, the magnetic field in this region of space becomes entangled too, and microseconds before particles disappear across the black hole's event horizon, to be lost forever, some can be captured by that magnetic field and funnelled away into space. These show up as jets.

In 1918, an American astronomer known as Herb Curtis (whom we shall meet properly in the next chapter) noticed that a nearby galaxy called M87 displayed a 'curiously straight jet' emanating from its central region.

The most powerful active galaxies generate more energy per second than a trillion Suns, with the result that the active nucleus outshines the rest of the galaxy by a hundred times or more. This brilliance masked the nature of an active galaxy for some time; when astronomers caught their first glimpses of them during the 1950s, they saw the star-like active cores and assumed that they were peculiar nearby stars. They called them 'quasi-stellar' objects, from which the present name quasar is derived.

The quasars were unmasked, however, in 1962 when astronomers discovered that they were located at great distances. That meant they could not be single stars but had to be extremely powerful galaxies, seen from a long way away.

Less powerful active galaxies can be found throughout the Universe at all distances. Some may be ageing quasars whose food source is almost used up. When the supermassive black hole finally devours everything within its reach, the active galaxy quietens to become a normal galaxy, such as our own. But there is nothing to stop the black hole coming back to life if more matter falls into its clutches.

According to calculations, one medium-sized star like the Sun wandering too close to the galactic centre is all that would be needed to reignite the activity and keep the black hole spewing energy for a year. So, the current population of active galaxies must be transient. If we came back in a million years' time, some presently active galaxies would have become inactive, while other, currently quiet ones, would be blazing with energy.

Astronomers had hoped for a flare-up from Sagittarius A* in spring 2014. They had been tracking a gas cloud near the centre of the Galaxy. With a disappointing lack of aplomb, they called it G2. Observations showed that it was being pulled to pieces by the gravity of Sagittarius A* and although the main bulk of the cloud would miss the supermassive black hole, they expected some of it to hit home. As it struck the accretion disc, there was every expectation that it would generate a blast of gamma rays to tell astronomers about the properties of both the gas cloud and the accretion disc.

The telescopes were trained. The time arrived. Nothing happened. As *Nature* magazine reported on its news pages in July 2014: 'Nothing. Nada. Zilch.' Everything missed the black hole.[†]

† www.nature.com/news/why-galactic-black-hole-fireworks-were-a-flop-1.15591

But astronomers have seen what appears to be matter falling into Cygnus X-1. The observations lay buried for a decade in the Hubble Space Telescope data archives until Joseph F. Dolan of NASA's Goddard Space Flight Center, Maryland, found the proverbial needle in the haystack.

On three separate occasions, back in June, July and August 1992, the Hubble Space Telescope was used to look at the stellar black hole and measure the brightness of its surrounding gas 100,000 times a second for an hour.

Far from the public favourite it is today, back then Hubble was an embarrassment. It had been launched in 1990 and was found to have defective optics. The 2.4-metre primary mirror had been polished to the wrong shape. To add to the shame, a NASA investigation showed that Perkin-Elmer, the company that manufactured the mirror, had noticed the flaw during testing and ignored it on the grounds that another, less-precise, test had failed to pick up the manufacturing error.

The result was that Hubble's images were ten times worse than they were supposed to be; barely were they an improvement on what could be taken from the ground at a fraction of the cost. As NASA scrambled to find a way to correct Hubble's vision, the only instrument that was relatively unaffected was the European-built photometer. Instead of creating pin-sharp images, all it did was measure brightness. So it was given priority to make observations until 1993, when a corrective optics unit would replace it to bring the cameras up to full muster.

The Cygnus X-1 observations totalled a billion data points. If printed out on a graph, the paper would stretch for 600 miles. Unsurprisingly, it took Dolan years to search the data. He was looking for pulses of ultraviolet radiation from hot clouds of gas that detach themselves from the inner edge of the accretion disc and begin their death spiral into the black hole. As they cascade in, the brightness fades in a way that is predictable from general relativity. The big test, however, was what happened at the end.

If the gas cloud were to hit something with a solid surface – say, a planet or a normal star – the impact would produce a spike of brightness. If it crossed the event horizon of a black hole, it would simply disappear from view. Dolan found two examples of such dying pulse trains in the data. One cloud had orbited the black hole six times before slipping into oblivion, the other seven times. In each case, it had taken the blobs just 0.2 seconds to orbit the black hole. Both had simply vanished rather than impacted.*

Now their atoms are part of one of the greatest mysteries facing astronomers: what lies inside a black hole? As we have mentioned, here we encounter something that cannot yet be answered even by our best theories. A black hole is not just a hole in the Universe; it is also a hole in our understanding of the Universe.

To make progress, we need to understand the behaviour of gravity on the smallest scale. This is known as quantum gravity and has become something of a holy grail for physicists. To gain some foothold into what a quantum theory of gravity might be, we need observations that show a deviation from general relativity, in the same way that Einstein used Mercury's recalcitrant orbit to leap beyond Newton.

Around the turn of the millennium, as if in answer to those

* Joseph F. Dolan, 'Dying Pulse Trains in Cygnus XR-1: Evidence for an Event Horizon?', *The Publications of the Astronomical Society of the Pacific*, Vol. 113, Issue 786, pp. 974–982.

prayers, spacecraft engineers found two subtle anomalies in the way their vehicles moved through space.

NASA launched Pioneer 10 and 11 on 2 March 1972 and 5 April 1973. They were the first robotic scouts sent to the outer Solar System, and relied on the fact that they were spinning to keep them stable, in the same way that a child's spinning top stays upright when it is rotating. The spin meant that the mission scientists hardly ever needed to fire the spacecraft's thrusters, and so for many months at a time, the Pioneers would be moving solely under the influence of the Solar System's gravitational field.

After they encountered the giant planets Jupiter and Saturn, returning the first close-up images and data from these worlds, they sailed on into the inky depths. Unwilling simply to terminate communications with two perfectly good spacecraft, NASA cast around for ideas of what to do with them.

John Anderson at NASA's Jet Propulsion Laboratory (JPL) and others came up with an objective. They would analyse the trajectory of the Pioneers in the hopes of finding a previously undiscovered planet in the outer Solar System. They reasoned that the gravity of such a planet would pull the little spacecraft off course by a minute but noticeable amount.

By the end of the decade, they had a signal. It was equivalent to a force 10 billion times weaker than Earth's gravity and it showed up in both spacecraft. But instead of pointing into the dark reaches of the outer Solar System, where they expected an undiscovered planet would lurk, the force pointed back towards the inner Solar System. It was as if the Sun's gravity were tugging both spacecraft a little bit harder than expected.

By the time NASA lost contact with Pioneer 10 in 2003, it was some 400,000 kilometres off course. To find out once and for all what lay at the base of this anomaly, in 2005 another JPL scientist, Slava Turyshev, decided to track down the telemetry data from the spacecraft. This had been generated by 114 on board sensors and would tell him the state of the spacecraft and its instruments. He would correlate it with the tracking data to see exactly how and when the Pioneers had gone off course. The original analysis couldn't determine whether it was changing with time or distance, whether it was always there or ramped up.

What Turyshev did, with a team of volunteers and some second-hand equipment that could read the decades-old data tapes, was virtually refly the mission inside a modern computer. This was no easy task. At the end of seven years' worth of effort to do this, Turyshev made his announcement. Disappointingly for those seeking new physics, it was all a red herring.

The Pioneers had been powered by small radioactive samples of plutonium. The virtual refly showed that the anomaly correlated with the stray heat generated by the isotope that escaped into space. Tucked behind the main antenna dish, which was always pointed back towards the inner Solar System, the pressure of the escaping infrared radiation hit the underside of the dish, effectively slowing down the spacecraft and making it look as if the Sun was pulling on it too harshly.

But as that avenue closed, another had opened up.

In 1990, NASA's latest Jupiter-bound spacecraft flew past the Earth. Such fly-bys are routinely used to gain speed and slingshot space probes towards the outer Solar System.

This time it was the turn of the Galileo craft. It shot by our planet, less than 1000 kilometres above the surface at a speed of 13.74 kilometres per second.

As NASA's tracking stations watched it head off, navigators noticed a small discrepancy in the radio signal. It could be explained only if the spacecraft had picked up about 4 millimetres per second more speed than they were expecting.

They investigated the navigation and tracking software but found no errors. The conclusion was stark: Newton's laws of gravity and motion could not account for the extra kick, so how had the spacecraft picked up the extra speed? Was this a tantalizing glimpse of unknown gravitational behaviour? If so, it could lead to a revolution in physics.

The energy kick became known as the fly-by anomaly, and as the Galileo probe came back for a second pass of Earth just two years later, so NASA readied its Tracking and Data Relay Satellite System. This time, the fly-by would be just 300 kilometres in altitude.

Disappointment prevailed. The fly-by was so close that Galileo flew through the last wisps of Earth's atmosphere. This was enough to slow down the spacecraft and mask any velocity increase comparable to the one it had received in 1990.

Time passed until on 28 January 1998, NASA's Near Earth Asteroid Rendezvous mission lined up for its slingshot, and emerged a whopping 13 millimetres per second faster than navigators were expecting.

In 2005, ESA's comet chaser Rosetta experienced a mysterious kick during the first of three planned Earth fly-bys. Like NASA with Galileo before it, ESA began to prepare for 2007, when Rosetta would be back for its second pass. They saw nothing, not even a hint. They tried again in 2009, during Rosetta's third and final pass. Again, they detected nothing.

'We were really hoping to see something on that third pass. We prepared very well and were really disappointed not to see anything,' said Trevor Morley, flight dynamics analyst at ESA's European Space Operations Centre, Darmstadt, Germany, in a telephone interview I conducted with him some years later. This marked perhaps the first time in history an engineer was disappointed because his spacecraft performed exactly as expected.

There had been more bad news when NASA's Mercury probe Messenger sailed past Earth in 2005 at 2336 kilometres. There was no unexpected gain in speed. But it was when Juno, NASA's latest Jupiter mission, passed Earth in October 2013 without manifesting an anomaly that Morley had his epiphany.

The key was Rosetta. If the fly-by anomaly really was the work of a previously undiscovered force of nature, then it would manifest itself time and again. The fact that Rosetta experienced it once but not a second or third time signalled to him that the cause must be on the spacecraft.

Unlike the early Pioneer spacecraft, Rosetta did not spin to keep stable. Instead, a sequence of cameras and thrusters kept its cubic body in a fixed orientation. This is known as three-axis stabilization and is the norm for most modern spacecraft. For Rosetta, two of the spacecraft's faces were designed almost never to face the Sun. Known as cold faces, this same design was used for two sister missions: Venus Express and Mars Express. Scanning the tracking data, Morley found that for all three craft there had been small changes of velocity on every rare occasion that sunlight had fallen onto a cold face.

His working hypothesis is that tiny quantities of moisture froze to these faces. When sunlight melted it, the gas shot from the surface, pushing the spacecraft in the opposite direction. In all cases, the velocity boost was a few millimetres per second – exactly within the range of the fly-by anomaly. The masterstroke was looking at the orientation of the spacecraft at the time of the fly-by. During Rosetta's first pass, it slipped between the Earth and the Sun, passing above the daytime hemisphere of our planet. The cold face was angled away from the Sun as usual, but that meant it was pointing at the Earth's day-lit side. The reflected sunlight was enough to cause whatever fluid had frozen there to outgas. When Rosetta returned for a second and third pass at higher altitude, there had not been enough time for any appreciable quantities of ice to re-form, so there was no fly-by anomaly.

Now, after a full decade in space, Rosetta still exhibits some outgassing when a normally cold face is exposed to the Sun. On 19 May 2014, after years in hibernation, the flight team were manoeuvring Rosetta and the Sun shone directly on the base plate of the spacecraft's lander. It clearly heated some ice that jetted off the spacecraft, because the tracking team measured a change of 0.5 millimetres per second in the spacecraft's velocity.

Satisfied with this explanation, the ESA has now shut the book on the fly-by anomaly. Once again, another possible chink in gravity's armour has been closed. But that doesn't stop astronomers looking for other weaknesses.

Tom Murphy did not exist when the experiment he now stewards began. He was born after the Apollo Moon landing missions, but now, every clear night when the Moon is high, the team he works on takes aim at our nearest celestial neighbour.

In a split second they blast it with 300 quadrillion particles of light from a powerful laser. They keep doing it, over and over again. Their targets are suitcase-sized reflectors left on the lunar surface by the Apollo 11 astronauts and two subsequent Apollo missions. Two Russian landers also carried reflectors on which the team can fix their sights. Out of every 300 quadrillion photons of light that the astronomers send to the Moon, just five find their way back to the waiting telescope on Earth. The rest are lost to the atmosphere of the Earth, either on the way out or the way back, or they miss the lunar reflectors.

From this small catch, the astronomers measure the movement of the Moon, looking for cracks in Einstein's general relativity. If they succeed, the Lunar Laser Ranging (LLR) experiment will become Apollo's greatest scientific legacy. But perhaps the most extraordinary part of this great experiment is that it partly owes its existence to the seamstresses of a tiny American town called Frederica.

With a total area of less than one square mile, and a population of about 750 people, Frederica, Delaware, seems an unlikely place to have played a major role in the exploration of the Universe. Yet it was here, back in the 1960s, that the spacesuits that went to the Moon were designed and made.

At the time, the company was part of Playtex, which made bras, girdles and other form-fitting undergarments for women. This made the ladies on the line perfect for stitching spacesuits that clung to the astronauts like second skins. Yet despite their expertise, the many layers necessary to keep a person alive so restricted movement that the astronauts found it difficult to do anything dextrous. The problem was discussed at a meeting of the Apollo Science Advisory Committee. One member was gravitational physicist Robert Dicke. He listened and then suggested that the astronauts simply set down some mirrors, angle them roughly at Earth and let astronomers back home do the rest. He calculated that with just a few years' worth of data, lunar laser ranging could test Einstein's principle of equivalence.

There is no getting around the fact that this principle sticks out like a sore thumb. Recall that it states acceleration is indistinguishable from the action of a gravitational field. It is bound up in the concept of an object's mass. In Newton's equations, mass is present in two places: once in his laws of motion and once in his law of gravity. The former refers to the inertia of a body – how hard it is to change its state of motion – and the latter to how a body responds in a gravitational field.

There is no reason why these two masses must be the same, yet to the limits of our experimental ability, the inertial mass and the gravitational mass appear identical. This is the most mystifying thing in the whole Universe. It must be telling us something profound. Einstein just accepted it and postulated the principle of equivalence, and general relativity follows – but only if these masses are perfectly equivalent.

If the masses are not identical, the principle of equivalence will not be strictly true and celestial objects will diverge from the motions predicted by relativity, much in the same way that Mercury's orbit diverged from Newton's predictions. The more massive the object, the more gravity it generates, and so the greater any deviation would be.

In the mid-1960s, Dicke and a colleague, Carl Brans, developed a rival theory to general relativity. By postulating a fifth force of nature, the Brans–Dicke theory of gravitation broke the equivalence principle and predicted a 13-metre perturbation in the Moon's orbit. Laser signals reflected from the Moon could prove the existence of such a disturbance.

So the astronauts placed the mirrors, and astronomers started measuring the position of the Moon. Sadly for Dicke, his theory became an early victim of the experiment because his 13-metre deviation was not seen. Nevertheless, new theoretical approaches to particle physics, such as string theory (see Chapter 7), and quintessence theories of dark energy (see Chapter 9), all imply that the equivalence principle must break. So, American funding agencies continue the experiments to this day, and they are getting more and more accurate all the time, thanks to upgrades of the laser system used.

Murphy and colleagues now measure the Moon's movement to an accuracy of a millimetre or two. At this scale, they can see that the pressure of the Sun's light pushes the Moon about 4 millimetres off its calculated path. They can see the expansion of the lunar surface as the Sun's rays heat it up every lunar morning, and they can watch it shrink again every lunar evening.

When they have factored out all these effects, they can compare the Moon's orbit with the predictions of general relativity. So far, they have seen no gravitational deviations whatsoever. The measurements are precise enough to show that gravitational mass and inertial mass are equivalent to an accuracy of one part in 10^{13}. This severely constrains how strong any possible fifth force of nature could be.

Others have also taken up the challenge. They test the equivalence principle in a different way by precisely measuring free-falling objects. This was apocryphally tested by Galileo dropping a bag of feathers and a bag of lead shot from the Leaning Tower of Pisa. Astronaut Dave Scott, whom we encountered in Chapter 2 when he and fellow Apollo 15 astronaut Jim Irwin found the Genesis Rock, did it for real using a hammer and feather while on the Moon. In the absence of air resistance, both objects hit the ground simultaneously.

But if the equivalence principle is wrong, then they could hit at slightly different times. The difference would likely be minute, but even the slightest difference would mean that general relativity is built on an approximation, and that would open the way for a more precise understanding to be found.

The University of Bremen's 146-metre-drop tower measures the fall of individual atoms of rubidium and potassium. Looming like a giant white rocket over the plains of northern Germany, it was inaugurated in 1990 as part of the Centre of Applied Space Technology and Microgravity (ZARM). So far, there has been no deviation from the behaviour predicted by the equivalence principle; the atoms have been found to fall at the same rate to accuracies of 11 decimal places.

Over at the University of Washington in Seattle, a team of researchers have built 'Eot-Wash', a high-tech set of scales known as a torsion balance, which can compare the motions of standard masses made of different elements, such as beryllium, aluminium, copper and silicon. They hold the record for accuracy for this method, with no violations of the equivalence principle to 13 decimal places.

More accuracy will require working in space, where the effect of Earth's gravity is diminished and deviations from the equivalence principle will be easier to spot. MICROSCOPE

(Micro-Satellite pour l'Observation Compensée à traînée du Principe d'Equivalence) is a French-led mission due to launch in 2016. It will test the motions of masses of platinum and iridium, with results 100 times more accurate than from any laboratory on Earth.

An even more sensitive mission, the Space-Time Explorer and Quantum Equivalence Principle Space Test is currently being evaluated by the European Space Agency, but so far the bare fact of the matter is that general relativity has passed every test that we have set it. That gives us a dilemma because, on the face it, it looks like general relativity is the ultimate gravitational theory, yet it can't explain black holes. If we believe that the Universe is understandable through mathematics, then black holes are telling us that there must be a deeper theory of gravity to be found. It's the intellectual equivalent of a rock and a hard place and the Universe is certainly not willing to give us much to work on.

And, as we'll now discover, from here on in the problems just get worse. Prepare to enter the truly unknown Universe.

6

The Luxuriant Garden

The eighteenth-century Hanoverian astronomer William Herschel looked at the night sky and saw not an unknown realm but a luxuriant garden. He lived in Bath, England, with his sister, Caroline. At a time when Royal Society President Joseph Banks and other botanists were travelling near and far to classify the world's plants, Herschel's own efforts were directed beyond the planet.

He built hundreds of telescopes, referring to each by the length of the tube,* and undertook a survey of stars in the 1770s using a telescope that was 7 feet in length, with a mirror 6.2 inches in diameter. Working with Caroline as his amanuensis, he found stars in orbit around each other (double stars), stars that varied their brightness (variable stars), and one star that wasn't a star at all. It was the planet Uranus.

Falling into Herschel's sight in 1781, the planet's discovery was an extraordinary moment for astronomy. The other planets had all been known since antiquity because they could be seen by the naked eye. Yet here was a new world, unanticipated and unimagined. Measurements of its orbit soon showed that it was about twice as far away as Saturn, placing it almost twenty times further from the Sun than the Earth. This greatly expanded the known extent of our Solar System.

* Today telescopes are referred to by the diameter of their mirror, as this provides an immediate feel for the amount of light they can gather.

With an eye for personal advancement, Herschel proposed naming the planet after the reigning monarch, astronomy enthusiast George III. Although the suggestion met with the same amount of enthusiasm from fellow astronomers as Halley's earlier attempt at naming a constellation for Charles II, it served its purpose. George III appointed Herschel the King's Astronomer. This was a singular title, distinct from the Astronomer Royal, who worked at Greenwich and was Nevil Maskelyne at the time.

The king established the Herschels in Datchet, close enough to Windsor for him to pop round uninvited with various dignitaries and state guests to look through the telescopes. He also began to fund Herschel's development of larger instruments.

A 20-foot-long telescope allowed William and Caroline to begin a new survey – not of the stars this time but of the nebulous objects that are glimpsed in between. These faint smudges were what Herschel likened to flowers in the luxuriant garden. The larger the telescope, the more they could find, cataloguing thousands in a list that is still used today. Rather ironically, given its centuries-old pedigree, it is still known as the New General Catalogue.

As Herschel built larger and larger telescopes, some of these smudges resolved into collections of stars while others remained steadfastly nebulous. Emboldened by his success, Herschel wanted to push his telescope design as far as possible and make the largest telescope in the world. He petitioned the king and won funding for a 40-foot-long telescope. It would be a cast-iron tube suspended between 50-foot-high wooden A-frames. The mirror would be 47 inches across, made of solid metal polished to gleaming reflectivity.

Having stumped up £4,000, the king took a keen interest in the construction. During one visit with the archbishop of Canterbury, the giant shape was lying on the ground next to the growing wooden frame. The king beckoned his guest to the mouth of the tube. 'Come, my Lord Bishop,' he said. 'I will show you the way to Heaven.'

The work proceeded. The frames were completed and the telescope hung between them. Then one chilly winter's night, with Orion rising above the eastern horizon, Herschel pointed the telescope at the misty patch just below the three stars of Orion's belt. He crawled into the tube, taking with him a lens, and began to search for the focal point at which the giant mirror would form an image. When he found it, the Orion Nebula jumped at him, brighter than ever and resolved into a knot of four stars surrounded by veils of silvery gas. What a sight!

If only the rest of the telescope's career had been that successful. In truth the massive structure was cumbersome and difficult to point with much accuracy. One of the half-ton mirrors was badly made, and Herschel soon found himself returning to the 20-footer. But he could hardly tell the king that the great telescope had been a folly, so he maintained the charade of using the giant, while really performing his research with the smaller telescope. Nevertheless, the 40-footer came to symbolize astronomy's ambition to see farther. It became the emblem of the fledgling Royal Astronomical Society in 1820, and appears on the letterhead and website to this day.

William's only son, John, did not share the family passion for astronomy. After a precocious attempt to become president of the Royal Society failed, Herschel Junior decamped to South Africa, where he extended his father's catalogue of nebulae to the southern hemisphere and then declared his astronomical work was done. Nevertheless, a residue of commitment must have lingered.

In the 1840s, William Parsons, the 3rd earl of Rosse, began to indulge a passion for astronomy, and Herschel urged him to revisit the dream of a giant telescope. Parsons was in a position to do this. He lived at Birr Castle, Parsonstown, Co. Offaly, Ireland, and possessed a considerable fortune.

Through trial and error he developed the techniques needed to cast large telescope mirrors. By 1845, he had cast and polished two 6-foot-wide mirrors. All he needed was a giant tube and supports to construct a telescope. Instead of a wooden frame, he built two enormous brick walls to support the telescope's tube and began to consult William Herschel's catalogue to determine which nebulae were really distant star clusters and which were truly gaseous objects.

Using the Leviathan of Parsonstown, as the telescope became known, the earl took a special interest in the gaseous nebulae, especially those that displayed a spiral structure. These had been noticed in 1700 by English polymath Thomas Wright. At the time, Wright had speculated that the swirl of dim light, known to astronomers as the Andromeda Nebula, was a far-distant collection of stars similar to our own Milky Way. This was revolutionary. Until then, astronomers had assumed that stars filled the Universe in some unending sea. To make this new idea distinct, the philosopher Immanuel Kant named it the 'island universe hypothesis' in 1755.

In the wake of Rosse's Leviathan, telescopes continued to become more powerful. With every new instrument, so more spiral nebulae were discovered. As the numbers ran from hundreds to thousands, then tens and hundreds of thousands, many astronomers felt that to treat them all as distant collections of billions of stars strained credibility to breaking point. So a different view emerged, one that can also be traced to Kant.

In Chapter 1 we learnt how he reasoned that the Sun's planets had formed from a rotating gas cloud. Perhaps, thought the astronomers of the later nineteenth and early twentieth centuries, the spiral nebulae were relatively nearby gas clouds falling together under the pull of gravity and destined to become systems of planets much like our own Solar System. On the face of it, this seemed perhaps more reasonable than the island universe hypothesis, and it became the dominant point of view.

The key to proving which was right was to find some way to measure distances to the spiral nebulae. That didn't happen until the 1920s, during a great rivalry between astronomers Harlow Shapley and fellow American Edwin Hubble. In the aftermath, astronomers uncovered a mystery that led to the invention of 'dark matter', a hypothesis that continues to shape modern thinking.

Harlow Shapely became an astronomer by accident. It was 1907. Upon discovering that the University of Missouri's journalism course was not yet open for students, he opened the prospectus with the intention of studying the first course in the catalogue. Not sure how to pronounce archaeology, he flipped to the next subject: astronomy.*

He soon made a name for himself in the classes. Not only was he quick to absorb facts and act upon them, he also gained

* Known for his tall tales, this may well be Shapley's equivalent of Newton's apple story.

a reputation for 'thinking'. At the conclusion of his degree he was offered a job at the Mount Wilson Observatory in California. This was the 1920s equivalent of NASA. It operated the largest telescope in the world, with a mirror fully 60 inches across. It was at the forefront of research and clearly the instrument to use by any astronomer of ambition.

Shapley's interest lay in measuring the distance to the stars. The only direct way to do this is through parallax, a technique that was described as far back at Galileo and Kepler in the early seventeenth century as a way of proving whether or not the Earth was moving through space. The idea is simple: measure the apparent positions of a star in the night sky six months apart and compare them. A difference in position will be apparent because Earth is now on the other side of its orbit and our perspective on that star has changed. The amount of this change gives an angle that can be used with some trigonometry to calculate the distance of the star. The closer the star to Earth, the larger this angle will be.

But as astronomers discovered, even the largest angles are minuscule. Had Galileo been able to measure the parallax with his telescopes he would have avoided his infamous trial. In the event, it was not until 205 years later, in 1838, that Friedrich Bessel succeeded in finally measuring this elusive stellar parameter. He did it for the star 61 Cygni, showing that it was 10.3 light years away, which is within 10 per cent of the modern accepted value. This is one of the closest stars, yet its parallax angle is just 0.314 arc seconds, where 1800 arc seconds are needed to span the full Moon.

Although telescopes had improved greatly since then, even with the Mount Wilson 60-inch telescope most stars were still too far away to display any measurable parallax.

Thankfully, at Harvard, one of the lady computers had made a breakthrough.

Henrietta Leavitt had been assiduously studying photographic plates of a relatively nearby collection of stars called the Small Magellanic Cloud. With Herschelian verve, she had catalogued the almost 2000 variable stars that she saw, singling out sixteen as different from the rest. The pattern in which a variable rises in brightness and then fades again is called its light curve, and these sixteen light curves all followed the same shape. They would rise swiftly in brightness before dimming in a more leisurely fashion. This particular class of variable star became known as the Cepheid variables, after the first of its kind to be identified, Delta Cephei.

Leavitt's masterstroke was to notice that the greater the star's maximum brightness, the longer it took to complete a cycle of variation. Turning this around, she realized that by measuring the variation period, the star's true luminosity could be calculated. This luminosity could then be compared to how bright it appeared, and the difference used to calculate how far away it resides.

With this as his basis, Shapely set about identifying Cepheid variables in a type of spherical star cluster known as a globular cluster. These compact collections of stars dot the night sky, and Shapley reasoned that they were the boundary markers of the Milky Way.

According to the wisdom of the time, our Galaxy was 30,000 light years across, but Shapley's survey and subsequent calculations suggested it was more like 300,000 light years across: a size that was beyond anyone's imagination. Shapley argued that something so big must surely be the dominant structure in the Universe. If so, he reasoned, the spiral nebulae must be gas clouds and so were probably the sites of forming star systems.

We now know that there were a few weak links in Shapley's calculations (the Milky Way is about 100,000 light years across), but his logic was impeccable. He was clearly destined for greatness and in 1919 a position of power opened before him.

After forty-two years in charge of the Harvard Observatory, Edward Pickering died at the age of seventy-two. Shapely fixed his sights on the appointment, but knew the odds were stacked against him. He wrote to canvas support from two people. The first was, understandably, his boss: George Ellery Hale. The second was, bizarrely, the leading candidate for the job: Henry Norris Russell.

The answers he received were perhaps not what he was expecting. Neither felt it would be a good move for him. Hale went so far as to chastise him for throwing his hat in the ring, implying that such appointments are bestowed rather than open for competition.* Although Shapley wrote back to both and declared himself no longer interested, a secret visit by a Harvard trustee convinced him otherwise. He began politicking again, albeit somewhat more subtly this time. Then an opportunity fell into his lap.

Unbeknown to Shapley, a council meeting of the National Academy of Sciences held late in 1919 had raised the possibility of having a public debate. Shapley's boss, Hale, had proposed two possible topics. The first was general relativity; the second was the nature of the spiral nebulae.

* It is often said that part of the reluctance to appoint Shapley was because of his age. He was in his mid-thirties and apparently deemed too young. It is not clear that this is the case, however, as Pickering had ascended to the directorship at the age of thirty-one.

Hale favoured relativity. The world was still agog at Eddington's successful eclipse expedition. But surviving letters written by the secretary of the academy clearly show his misgivings: 'I must confess that I would rather have a subject in which there would be a half dozen members of the Academy competent enough to understand at least a few words of what the speakers were saying if we had a symposium upon it. I pray to God that the progress of science will send relativity to some region of space beyond the fourth dimension, from whence it may never return to plague us.'*

Instead, the secretary favoured a debate about the nature of spiral nebulae and Kant's island universe hypothesis.

Shapley was an obvious participant because of his work on the size of the Milky Way. He was invited to take part and informed that the debate would be held in Washington DC, close enough to attract the trustees of Harvard. This was his chance to impress them, but it was a gamble.

His opponent was Heber Curtis, an accomplished astronomer and public speaker from the Lick Observatory, San Jose, California. Two years earlier Curtis had seen a star explode, apparently embedded in the Andromeda Nebula. Astronomers were getting good at identifying these novae, as the bright explosions were known, but something about the one in Andromeda struck Curtis as puzzling. It was much fainter than many of the other novae that had been observed dotted along the Milky Way.

He inspected the photographic plates in the Lick library

* A detailed history of the 'Great Debate' can be found in The 'Great Debate': What Really Happened, Michael A. Hoskin, *Journal for the History of Astronomy*, 7, 169-182, 1976. (http://apod.nasa.gov/diamond_jubilee/1920/cs_real.html) More articles about it can be found at http://apod.nasa.gov/diamond_jubilee/debate20.html.

and found ten others. Six occurred along the Milky Way, four didn't. The ones that didn't were all about ten times fainter than those in the Milky Way and – tellingly – all fell within spiral nebulae. Curtis pointed out that ten times fainter could mean a hundred times further away, and if so, would support the island universe hypothesis.

Astronomers were reluctant to agree. Curtis had assumed that the diminishment was due to distance but he could not rule out that the novae had just been intrinsically faint. So the debate was conceived to look at the evidence. Curtis would present his side of the argument, and Shapley would talk about his.

The trouble was that Shapley knew Curtis could outclass him in the debating stakes. If he wasn't careful, he could be made to look like a tongue-tied buffoon – and who would want someone like that running Harvard's prestigious observatory?

Shapley began writing letters, and over a period of weeks, he managed to change the rules of engagement. Instead of a back-and-forth debate, it would be a discussion, with each side presenting a forty-minute lecture. Then a question and answer session could provide the debating arena, as much for the audience of astronomers as for the speakers themselves.

Shapley went first, and read his presentation word for word from extensive notes. He opened with such elementary astronomical material that it was beyond reproach. Then he subtly framed his presentation of the size of the Milky Way to make it look as if Curtis was a stick-in-the-mud by resisting the revised size of the Milky Way. It was a cunning ploy to insinuate that Curtis was being driven into exotic new hypotheses by his refusal to revise a basic parameter.

Shapley placed the emphasis of his talk on the bedrock of astronomy's understanding of the Universe, namely distance measurements – not on the details that sprang from it. The final section of his talk was about the development of new technology that would allow better observations to be made in the future.

Curtis came next, talking freely from a set of slides. He had marshalled an impressive set of facts and built his arguments carefully, but the nature of his talk was largely speculative because he still lacked a definitive observation that spiral nebulae were external galaxies.

In the ensuing discussion, Curtis dominated with his oratory skills. Shapley stumbled and was then put to shame by Russell, who stood up in the audience and delivered a rather more eloquent version of the arguments that Shapley had been trying to advance.

Curtis was judged the 'winner' and was soon referring to the meeting as 'our memorable set-to'. By 1926, an article in *Popular Astronomy* magazine called 'Do We Live in a Spiral Nebula?'* referred back to the 'memorable discussion' of 1920. And gradually the myth of the great debate gained momentum. But what did Curtis actually win? Nature cannot be decided by debate. What we think has no currency. The world and the wider Universe are simply what they are.

For Shapley, the debate was a disaster. The scout from Harvard reported that he was talented but too immature to be offered the top job. So they offered Russell the directorship

* The crux of this article was that the Universe could be much larger than anyone anticipated. Here we are a century later, with people like me writing articles such as 'Do We Live in a Multiverse?', which are essentially another discussion about whether the Universe is much larger than we thought. *Plus ça change ...*

and Shapley a second-in-command slot. Russell agonized. He could see that the pairing would be magnificent; what the two of them didn't know about stellar astronomy was hardly worth knowing, but the directorship would necessitate him essentially giving up his own active research. He would have to content himself with directing others. When Princeton discovered that Russell was considering leaving them, they offered him an attractive new deal to stay and practise all the astronomy he wanted. The die was cast. Russell turned down the directorship at Harvard.

Shapley turned down Harvard too; he was not interested in leaving the largest telescope in the world for anything other than the leadership. It was his boss, Hale, who came to the rescue. Previously opposed to Shapley's ambition, he suddenly suggested a compromise: Harvard could take on the young man in a trial run. If he succeeded in his first twelve months, then he should be confirmed for the job on a permanent basis.

Meanwhile, in the wider astronomical community, the nature of the spiral nebulae continued to spark interest. What was needed to settle the issue was a clear measurement of distance. In practice that meant someone had to identify a Cepheid variable in one so that its distance could be calculated. Beyond 300,000 light years and the spiral nebula had to be a galaxy in its own right. Had Shapley stayed at Mount Wilson, he could perhaps have been in a position to do this work. And if he had, he would now have the most famous telescope in the world named after him. As it was, many decades later the Hubble Space Telescope was named after his great rival, Edwin Hubble.

• • •

Hubble, too, was born in Missouri, four years junior to Shapley. For his eighth birthday, he was allowed to stay up past his bedtime – in itself not an uncommon childhood gift – but it was what he did with it that set the course of his life.

Hubble's grandfather had built a telescope, and on that night, the young boy was allowed to look through it for the first time. Forever afterwards, he wanted to be an astronomer. His father had other ideas and when Hubble entered university it was to study the prerequisite courses for a career in law. In his spare time, he took classes in mathematics, astronomy, and the other sciences.

He was a driven individual, being chosen in 1910 as the Rhodes Scholar for Illinois. His reward was three years of paid study at Oxford University, England. There he again enrolled in law, and began the most extraordinary personal transformation. He shed his American trappings, adopting a fake English accent that entailed shouting 'By Jove!' at the top of his voice in response to matters of little significance. He learnt to wear a cloak and became forever 'gasping for a cuppa'. His affected identity did not impress everyone, but he worked hard and a career in law seemed inevitable.

But Oxford was redolent of astronomy: the great Edmond Halley had lived and worked there. His house was just two minutes from the university's exquisite ceremonial hall: the Sheldonian Theatre. Also, in the year that Hubble arrived at Oxford, Halley's comet had returned to put on a dazzling display in the night sky and, secretly, Hubble had begun spending time with the astronomers. Then, in February 1913, devastating news arrived.

Hubble's father had died, taken by kidney disease.

At the completion of his studies in May, Hubble returned

to the States and slowly the implications of losing his father sank in. Although he wrote to his friends in Oxford of completing his final law examinations, he was in reality teaching physics and mathematics at a high school in New Albany, Indiana.

The truth was that without his father's anchoring presence, the Universe was free to draw his attention upwards. By autumn 1914, he was employed at the University of Chicago's Yerkes Observatory to work on his thesis, the nature of the spiral nebulae. He all but commandeered one of the smaller telescopes and began photographing the beautiful celestial whirlpools.

Following a sojourn in the army upon America's entry into the First World War, Hubble moved to Mount Wilson under the supervision of Hale. Here, he had access to the largest telescopes in the world: the 60-inch and the newly commissioned 100-inch telescope. His first night's observing took place on 18 October 1919.

At the observatory, Hubble and Shapley met – and circled each other like cats. Hubble drifted around the observatory aloof from his peers. He had taken to wearing the most extraordinary observing uniform: jodhpurs, leather puttees and a beret. Meanwhile Shapley exuded the kind of loquacious southern charm that Hubble despised and the rivalry was established.

Once Shapley left for Harvard, Hubble forged ahead. As Shapley knuckled down to proving what a good administrator he could be, Hubble continued his assault on the nebulae. He turned up night after night with a careful plan of observation, and on 4 October 1923 he turned towards Andromeda.

His photograph showed that the misty spiral now displayed three individual stars. His first assumption was that these were novae. To check whether these particular ones had been seen before, he went to the observatory's plate library and began checking previous photographs.

Two of the new stars were indeed nova explosions; they had never been seen before. But the third one was special. It sometimes appeared on previous images, other times not. It was a variable star. Hubble crossed out the 'N' he had written on the most recent plate and excitedly wrote 'VAR!' Could it be – dare he hope that it could be – a Cepheid variable? If so, he could measure its distance.

He returned to photograph Andromeda night after night. In early February 1924, he watched the star take a precipitous upturn in brightness. With months of data in hand, he felt confident enough to construct a light curve. The heart-stopping moment was when he realized that the light curve rose sharply, and then tapered gradually. It was the shape of a Cepheid variable with a period of 31.4 days. Turning to the published calibrations that Shapely had used to measure distances to the globular clusters, Hubble's jaw must have dropped. A variation period of just over a month signalled that the Cepheid's luminosity must be about 7000 times brighter than the Sun. It was a behemoth of a star, yet the dimmest minnow in the plates, appearing about 100,000 times fainter than could be seen with the naked eye.

Fastidiously applying Shapley's own method, Hubble calculated the distance at 1 million light years – more than three times further than Shapley's already giant size of the Milky Way. They could be no doubt: Andromeda was a distant galaxy in its own right. And by implication so were the hundreds of thousands of other spiral nebulae. The island universe hypothesis was true and the Universe was unimaginably larger than anyone had thought before. Hubble could not resist. He put pen to paper immediately: 'Dear Shapley…'

When the letter arrived, complete with graph and calculation, Shapley was in conversation with one of the lady astronomers. She remembered he held out the missive and told her, 'Here is the letter that has destroyed my universe.'

But Shapley had the last word over the naming of the spiral nebulae. Astronomers began to bandy around terms such as island universes, extragalactic nebulae, and the outlandish anagalactic nebulae. Shapley cut through it all. Claiming that since these objects were demonstrably not nebulae, nor universes, he was going to adopt the term 'galaxies'. This made perfect sense to him; galaxy was derived from the Greek *galaxias*, meaning 'milky', which is how the name Milky Way had been transformed into Galaxy.

The reasoning was that if the Earth possessed the Moon, but Jupiter and other planets possessed moons, why should we not live in a Galaxy, with other galaxies cast across the oceans of space, which could now be called intergalactic space? Hubble never accepted the term, doggedly referring to them as extragalactic nebulae until his death in 1956, when the term died with him.

The galaxies became the focus of attention for astronomers. They were mapped, catalogued and studied for the secrets of their composition and movement. It took just over a decade for a Swiss émigré astronomer with a penchant for rudeness to find a major problem in our understanding of the way galaxies move. The problem remains with us today, and although it has consumed billions in research funding it has arguably made little true progress.

Swiss contrarian Fritz Zwicky was known as a loudmouth. He claimed simply to 'call a child by its name' and considered 'humbleness was a lie', but most just considered him rude. He and his wife would routinely – and without insult intended – refer to Zwicky's students as bastards. One day, by invitation, his coterie turned up at the house for Sunday lunch. As Mrs Zwicky opened the door, she turned into the hallway and called cheerily, 'The bastards are here.' If Zwicky did want to insult then he used the term 'spherical bastard' because the person in question was a bastard 'when looked at from any side'.

He was renowned for standing up in the middle of research lectures to inform the speaker that the problem had already been solved – by Zwicky himself. In this respect he was the intellectual reincarnation of Robert Hooke. The irony is that today Zwicky is best remembered not for the problems he did solve, but for the one he couldn't.

In the immediate aftermath of the galaxies being identified as distant collections of stars, a natural question to ask was: how are they distributed through space? Hubble thought that the distribution was random and this let Shapley score against him. Together with Adelaide Ames, one of the Harvard computers, Shapley conducted a photographic survey of the northern hemisphere that showed galaxies cluster together. It implied they were gravitationally bound to one another.

This piqued Zwicky's attention. He was working at the Californian Institute of Technology and began using a recently completed photographic telescope at Mount Wilson to study the motion of galaxies in a collection known as the Coma Cluster. He counted a membership of about a thousand galaxies. Having measured their motion, he then estimated how much matter was needed to keep them moving. It was a surprisingly large amount, so he estimated the cluster's mass in a different way. From the brightness of the galaxies, he could estimate the mass of their stars. This time the figure was comparatively small: about 500 times lower than those based on the gravitational estimate.

He published his conclusions in German first, then in English in 1937 in the *Astrophysical Journal.* To resolve the disagreement in the estimates, he proposed that there had to be more matter in the galaxy cluster than could be seen in the images taken. In the German version of the paper, he called this *dunkle materie*. In translation: 'dark matter'.*

The bastards had arrived.

Zwicky thought dark matter had to be gas clouds that had yet to collapse into stars. And so, he spent uncountable hours at the eyepiece looking for the silhouette or the telltale spectral traces of these phantom nebulae. He came up empty-handed.

It was a puzzle for sure but not one that many astronomers were willing to devote much time to solving. Yet as the years and then the decades wore on, more and more astronomers were coming to the conclusion that there was 'missing matter' in the galaxies.

In 1959, a consideration of the Andromeda Galaxy showed that the speed at which it was approaching the Milky Way suggested these two mighty galaxies were pulling each other with a gravitational force hundreds of times stronger than could be explained by the mass of their stars. Dust clouds

* Zwicky, F. (1933), 'Die Rotverschiebung von extragalaktischen Nebeln', *Helvetica Physica Acta* 6: 110–127, and Zwicky, F. (1937), 'On the Masses of Nebulae and of Clusters of Nebulae', *Astrophysical Journal* 86: 217.

were known in many galaxies by this time, but none was overwhelmingly large. At most they might double the amount of matter but not boost it hundreds of times.

What was needed was another independent way to deduce the mass of a galaxy and show whether the 'missing mass' was a real problem or a chimera produced by inaccurate observations.

Since Newton, astronomers have known that the orbital speed of a planet is determined by the mass of the star it circles and its distance from that star. In 1970, Vera Rubin and Kent Ford extended the logic to look at the orbital speed of giant glowing gas clouds in the Andromeda Galaxy. Those on the edge of the galaxy would give them a good measure of the celestial object's total mass.

To do the work, they needed to take spectra of the clouds and then use the Doppler shift to calculate the motions. Rubin and Ford targeted the gas clouds using the 84-inch telescope at Kitt Peak. Night after night they worked in nearly complete darkness at the telescope. Each would take their turn at the eyepiece, staring at the galaxy for 70-minute stretches at a time, guiding the telescope with a hand control as the photographic plate gradually registered an image from the faint light.

As the nights dragged on, Rubin found the ghostly glow of the galaxy in the eyepiece spooky yet exhilarating. When the darkness became too disorienting, she would briefly flash her faint red torch, just to see an impression of the barrel of the scope and the walls of the observatory. Even so, her mind used to wander to thoughts of whether there was an alien astronomer in the Andromeda Galaxy looking back at the

Milky Way, performing a similar observation to her own.*

It took three years for them to collect their data. At the end of that time, they had measurements of sixty-seven gas clouds across the width of Andromeda. When they plotted the velocities on a graph, they realized that they had two problems, not just one.

First, they confirmed that there was not enough visible matter in a galaxy to produce the orbital velocities; the missing mass problem was real. The second problem they uncovered was that the outer gas clouds were moving more quickly than expected.

It was assumed that the outer clouds would be moving more slowly than the inner clouds, in the same way that the outermost planets move less quickly than the inner ones, but this was not the case. The peripheral clouds were moving just as quickly as the central ones, which was manifestly odd and went against everything that was understood about gravity weakening with distance.

Subsequent observations showed that other spiral galaxies displayed these 'flat rotation curves' also. To reproduce such orbits, astronomers turned to computer models. Since the distribution of visible matter was fixed by the stars that they could see, they began experimenting with dark matter to provide an extra pull of gravity. They found that to produce a flat rotation curve, a galaxy must be surrounded by a vast cloud of matter. This spherical cloud was named the halo and contained from ten to a hundred times more mass than the stars.

It was enough to convince most astronomers that there was a vast quantity of unseen matter that needed discovering

* www.ifsc.usp.br/~hoyos/courses/2012/FCM0102/2006_Physics_Today_Vera_Rubin_vol59no12p8_9.pdf

in the Universe, but the precise nature of this stuff proved tricky to pin down.

As we shall see in Chapter 8, cosmologists had focused on a scenario for the beginning of the Universe that became known as the Big Bang. In this cataclysmic event, all matter and energy burst into existence. That meant every atom in the modern Universe had its origin in the Big Bang. Calculations to work out how many atoms emerged from the Big Bang, and the relative proportion of the chemical elements, worked out just fine based on the amount of matter that the astronomers could see in the stars. If they added in the amount of atoms needed to make up the contingent of dark matter, however, then the sums went to pot.

And if all that was somewhat esoteric, then there was another, far simpler constraint. The amount of atoms that would be needed to fill the halo would absorb so much light that the galaxy inside would be rendered virtually invisible.

Every indication was that, whatever this dark matter was, it was not made up of atoms. So what was it? One physicist had an answer, but not many were going to like it.

In the early 1980s, physicist Mordehai Milgrom was dividing his time between the Weizmann Institute in Rehovot, Israel, and the Institute for Advanced Study in Princeton. He was as intrigued by the flat rotation curves as the astronomers, but perhaps because his background was physics rather than astronomy he saw things differently.

He wondered whether the force of gravity itself was changing in the distant reaches of a galaxy. Newton's law described how gravity diminished with distance in accordance with an inverse square law: double the distance, quarter the intensity. But the flat rotation curves were showing that this was not happening in a galaxy. It was as if the galaxy was pulling a little harder in those distant realms.

At first, Milgrom thought that there must be some sort of length scale over which gravity departed from a strict inverse square, but try as he might, he could not make the sums reproduce the observations. Then inspiration struck. What if it were not a fixed distance? What if there were a limit to the strength of gravity itself? In this case, the inverse square law would change once the force dropped below a particular strength. Since the effect of a gravitational force is to accelerate objects, Milgrom postulated a new constant of nature: an acceleration. Below this critical acceleration, the inverse square decrease of gravity changed to fall away more slowly. The effect of this would be that gravity was stronger in such weak fields than Newton's law predicted. He called his idea modified Newtonian dynamics, or MOND for short.*

He spent six months working to finesse his ideas, finding that the best fit to the observations took place when the critical acceleration was a hundred billionths of a g-force. That is roughly the gravitational field produced by the mass of a single sheet of paper in free space.

It is so small that most would have considered it negligible, but Milgrom showed that operating over cosmic distances it could speed up the stars in the outer reaches of a galaxy and produce flat rotation curves that looked remarkably like the ones being observed by astronomers. He published three papers in 1983† to present his ground-breaking idea to the astronomical community, and waited for a reaction. By his own admission, he was met mostly with silence.

* arXiv:astro-ph/0112069v1

† See www.astro.umd.edu/~ssm/mond/literature.html for references.

The big trouble was that he could describe the rotation of galaxies beautifully, but he could not say why this change in gravity should take place. Without an underlying theory, most astronomers were highly sceptical. If Milgrom were right, it meant that a whole new theory of gravity was needed – but not in the way that the black hole theorists were working towards.

As we saw in the last chapter, black holes are telling us that there is a need for a better theory when gravity is exceptionally strong. Milgrom's ideas meant altering gravity at the other end of the spectrum, in weak fields, and the very thought was too much for most theorists. No one seemed to have the inspiration or the stomach to contemplate another root-and-branch modification to fundamental physics. They were having enough problems extending gravity to account for black holes. The only option left was to return to the concept of dark matter, not that it was an easy path to take.

Even today there is not enough computing power to track the motion of every star in a galaxy, let alone every subatomic particle. Computer models are only capable of simulating the movement of large volumes of matter containing some 10,000 times the mass of the Sun. As a result, they are utterly incapable of predicting the behaviour of individual dark matter particles and the astronomers could make little progress.

Then the particle physicists had a suggestion.

7
Chiaroscuro

In the latter years of the nineteenth century, it was widely thought that atoms were the smallest pieces of matter, even though no one knew what an atom really was. English physicist J. J. Thomson shattered that view in 1897 during his investigation of cathode rays. These were recently discovered emissions that could be generated by a pair of electrodes. They were invisible until they struck the end of the glass vacuum tube in which they were generated, giving off a ghostly glow.

Thomson found that the rays could be deflected by magnetic and electrical fields. When he compared the amount of this deflection to that of electrically charged hydrogen atoms, the cathode rays behaved as if they were made up of particles approximately 1800 times lighter. Since hydrogen is the lightest atom, Thomson concluded that the cathode particles must be constituents of atoms. He termed them 'corpuscles' – although soon they became known as electrons – and erroneously proposed that atoms were rather like a plum pudding. The electrons carried the negative electrical charge and were the plums. The positive charge was the surrounding pudding, weakly spread throughout the rest of the atom.

The Japanese physicist Hantaro Hagaoka, who worked at Tokyo University, proposed a different – and as it would turn out, correct – model. Taking his inspiration from the rings of Saturn, he thought that the electrons were orbiting a compact nucleus of dense positive charge. The particles that band together to form the nucleus were called protons. In 1909, working at the University of Manchester, New Zealand-born physicist Ernest Rutherford found the experimental evidence to prove Hagaoka's view.

Already a Nobel laureate in chemistry thanks to his work with radioactive substances, Rutherford actually favoured the plum pudding model. Together with two junior researchers, Hans Geiger (who went on to invent the radiation detector known as the Geiger counter) and undergraduate Ernest Marsden, he set out to prove it.

One of them, usually Marsden, would sit in a darkened room for twenty minutes. In the almost total gloom, his eyes would gradually adjust to the dark. Feeling his way, he moved to the apparatus and looked through a microscope that pointed horizontally towards the experiment. Almost at once, Marsden began to see fleeting pinpricks of light through the microscope. This was strange. According to Rutherford, he was supposed to see nothing at all.*

The apparatus consisted of a sheet of gold foil, just a few hundred atoms thick, and a sample of radium bromide. This radioactive substance spat out positively charged debris called alpha particles. If the plum pudding model was correct, these should pass straight through the gold foil, because the positive charge there was spread too thinly to have much effect, but Marsden was seeing alpha particles bouncing almost straight back at him. They had been rendered into sparks of light by a fluorescent screen placed between the microscope and the foil.

* For a full treatment of these events, see *The Fly in the Cathedral* by Brian Cathcart, Farrar Straus Giroux

Scared of making a mistake, he repeated the experiment obsessively, watching the minute fire of subatomic events unfold for a week. He varied the set-up to see if there were some mistake in the way it had been constructed but nothing he did eliminated the flashes. Whatever the atom was, it was not a plum pudding. He told Rutherford and scientific history was made.

Years later, the New Zealander remembered that the news was 'quite the most incredible event that has ever happened to me in my life'. One can just imagine his bushy walrus moustache quivering as he fought for an analogy to bring the discovery to life. Eventually, he settled on: 'It was almost as incredible as if you fired a 15-inch shell at a piece of tissue paper and it came back and hit you.'

Eighteen months of concentrated effort began in which the three men ran more and more experiments. They varied the metal in the foil and the angle at which they observed through the microscope. Painstakingly, they realized that each atom had a small, dense, positively charged nucleus, as Hagaoka had proposed in his Saturn model. Most of the atom's volume is empty space in which the much lighter, negatively charged electrons can be found. Rutherford referred to the nucleus as being like a gnat in the Albert Hall, a phrase that has metamorphosed by usage into 'the fly in the cathedral'.

In tandem, theorists were constructing an extraordinary framework in which to understand the behaviour of this subatomic realm. As mentioned in chapter 3, quantum mechanics is the equivalent of Newton's laws of motion but for particles.

A cornerstone of this approach is that all matter, energy and forces are carried on particles. Indeed, one of Einstein's *annus mirabilis* papers had shown that light rays could be thought of as streams of particles called photons, each one a little packet of energy. The surprise was that in developing quantum explanations for the forces at work inside an atomic nucleus, theorists developed the ability to predict previously unanticipated particles. The person who took this to a dizzying new height, and paved the way for our modern assumptions about dark matter, was Wolfgang Pauli.

In 1930, Pauli was having trouble with his sums. Try as he might, he couldn't get them to balance. He was working on a theory of radioactivity, but something was missing from the calculations. It was as if energy were stealing away from the reaction unaccounted for.

When he identified a possible solution, he was so taken aback that he wrote to fellow researchers almost begging them to disagree with him. Addressing them as 'Dear radioactive ladies and gentlemen', he told them about his 'desperate remedy', which was to suppose that there must be an unknown particle of nature involved in the reaction, and that this was carrying away the missing energy.

The type of radioactivity he was investigating is known as beta decay. It makes radiocarbon dating possible, although at the time Pauli was simply trying to understand the process by which a nucleus spits out an electron to transform itself into another element.

According to Pauli's calculations, the electrons should be sent packing with a set amount of energy predicted by $E = mc^2$. The Geiger counters, on the contrary, were showing that electrons always carried less than the full amount.

Pauli's desperate remedy was to suggest that an unseen particle was carrying away the deficit. He named the newbie the neutron and made a stab at deducing its properties. It was electrically neutral – hence his suggested name – and he thought it was roughly the same mass as an electron. What frustrated Pauli, however, was the belief that the experimentalists should have already discovered it if it really existed.

Recognizing that his reputation was on the line, he quipped in the letter that 'only those who wager can win'. The 'radioactives' started thinking about the new particle and in 1932 it looked as if the breakthrough had been made. Working at the University of Cambridge, James Chadwick discovered an electrically neutral particle that was emitted under certain conditions from atoms. However, as investigations soon showed, this 'neutron' was much too heavy to be the particle that Pauli had suggested. It was still a major discovery, just not the one physicists were expecting. Then a new clue emerged.

Just a year later, Italian physicist Enrico Fermi took inspiration from Pauli's suggested particle, and used it to develop a fuller mathematical theory of beta decay. The twist was that, in order to work, the theory needed a new force of nature to be working inside the atom.

Forces lie at the heart of physics. They are the means by which matter interacts with itself. As we have discussed, gravity was the first to be tackled by physicists, electromagnetism was the second. Whereas gravity only attracts, electromagnetism has polarity: positive and negative charges for electricity, north and south poles for magnetism. Like polarities repel, unlike polarities attract. This is why plugs won't work if you wire them incorrectly, and why magnets stick together one way but resist each other if turned around.

By the 1930s, it was clear that there had to be another force at work inside the atomic nucleus. The protons, being positively charged, would repel each other. Without something to overwhelm this repulsion, the nucleus would blow itself apart. Chadwick's neutrons were no help because they were electrically neutral, so there had to be another force of nature acting strongly between the protons and the neutrons to hold everything tightly. With the kind of creative flare only physicists can manage, they called it the strong force.

Initially, it looked as if Fermi's proposed force might fit the bill, but based on the observed rates of beta decay, his calculations soon showed that the strength was far too weak to hold the nucleus together: clearly there were two forces at work in the nucleus, the strong force to hold it together and Fermi's weak force to drive some forms of radioactivity – but only if Pauli's postulated neutral particle actually existed. Academically, the stakes could not have been higher. To distinguish it from Chadwick's neutron, Fermi named it the neutrino, which in Italian means 'little neutral one'.

Unfortunately, progress was not rapid. The English-speaking community of scientists had to wait six years from the initial publication of Fermi's ideas before being able to read them in their native language. This was because leading science journal *Nature* had initially rejected the paper on the grounds of it being too speculative. Hence, Fermi had published it in French and German journals. Only in 1939 did *Nature* relent and publish an English-language version. By then, interest in the particle and the theory behind it was gaining momentum, but the task of building a neutrino detector seemed highly impractical, if not impossible.

It was calculated that a neutrino interacts so rarely with matter that a sheet of lead more than a light year deep would be needed to guarantee stopping it. In consequence, the mighty Earth itself would appear virtually transparent to neutrinos. Faced with such statistics, the particle physicists turned to more pressing concerns. Uranium had been discovered to be capable of sustaining a chain reaction, and that made it conceivable to build an atomic bomb. Weeks before the outbreak of the Second World War, Albert Einstein signed a letter to President Roosevelt alerting him to the threat of Nazi Germany developing such a weapon.

By 1940, it was realized that the amounts of uranium needed for a bomb were much smaller than previously thought and the allies initiated the Manhattan Project at Los Alamos Scientific Laboratory, New Mexico, to make the doomsday device. By the summer of 1945, they were ready. On 6 and 9 August, Hiroshima and Nagasaki fell victim and witness to the power of the atom.

Astonishingly, these terrible weapons had been developed without a proven theory of how the atom stayed together. A full mathematical treatment of the strong nuclear force did not arrive until the 1970s. The Manhattan Project physicists had done it all with experimental trial and error rather than from first principles. Who needed the theoretical purity that James Jeans had championed just a few decades before when such extraordinary powers could be harnessed or released by application alone?

Nuclear reactors had been a by-product of the atomic bomb. Fermi had fled Italy in 1938 to protect his Jewish wife against prejudice, and in 1942 led a group at University of Chicago to produce the world's first artificial reactor, Chicago Pile-1. It was this success that led to the experiment that proved the reality of his hypothesized neutrino particle.

According to his hypothesis, the reactors he pioneered would be neutrino factories producing billions every seconds. But it didn't occur to anyone straight away to see whether these could be detected. That job fell to Captain Clyde Cowan of the US Army Air Forces.

Cowan left active duty after the Second World War in 1946 and was awarded money from the government's Servicemen's Readjustment Act to study for a masters degree and then a PhD at Washington University in St. Louis, Missouri. From there he joined the staff at Los Alamos and met Frederick Reines, who had also worked on the Manhattan Project. In the spirit of optimism that permeated the United States following the war, they identified a mutual ambition to tackle some 'challenging physics', and it didn't come any more challenging than trying to detect the neutrino.

According to Fermi's theory, if a neutrino were to hit a proton, it would transform that particle into a neutron and a pair of gamma rays that carried a set amount of energy, given by $E = mc^2$. All they needed was a target rich in protons and a source of neutrinos.

For the former, they used a tank of water. Hydrogen's atomic nucleus is a single proton, and with two hydrogens in every molecule of water, it was a rich source. Next, they needed a source of neutrinos. They considered setting up their experiment near to a nuclear explosion but then remembered the nuclear reactors.

As part of the Manhattan Project, a reactor had been built in Washington State on the banks of the Columbia River. Known as the Hanford Site, the reactor had been used to produce the plutonium that destroyed Nagasaki.

Cowan and Reines calculated that it was producing neutrinos at an astonishing rate: 10,000 billion per second were passing through every square centimetre that surrounded the reactor. This would compensate for the highly unreactive nature of the neutrinos. So they set up their 200-litre water tank, surrounded it with an array of gamma-ray detectors and switched it on. They were overwhelmed with signals – far too many to be from neutrino interactions.

The trouble was that the radiation from the reactor kept fooling the detector, washing out the neutrino signals. After two years of trying to fix the problem, they conceded defeat and moved the experiment to the newly opened Savannah River reactors in South Carolina. Here they could set up in a space 12 metres underground from reactor P. This provided the much-needed shielding to block out the unwanted radiation and within a year they had all the gamma rays they needed to claim detection. They announced the discovery of the elusive particle in the journal *Science*.* It opened whole new realms of particle physics and astronomy, and a way to make a final conclusive test of Eddington's model that the Sun derived its energy from atomic reactions.

Back in 1926, as discussed in Chapter 3, Eddington had suggested that hydrogen transforming into helium powered the Sun. By 1939, physicists had turned that into a complete theory of the reaction, and it relied on neutrinos. Known as the proton–proton chain, the reaction predicted that a neutrino would be released in the first step of the chain. Unlike light, which instantly became bogged down in the dense gas of

* 'Detection of the Free Neutrino: A Confirmation', Cowan, C.L., Jr.; Reines, F.; Harrison, F.B.; Kruse, H.W.; McGuire, A. D., *Science*, Vol. 124, Issue 3212, pp. 103-104.

the Sun's core, neutrinos would leave immediately. Scientists realized that, just as Cowan and Reines had put a neutrino detector next to a nuclear reactor to capture the neutrinos generated inside, a suitably large detector on Earth would be 'next door' to the reactor of the Sun.

It took until the mid-1960s for physicists to make such a technological beast. By then, the proton–proton chain had been widely studied by laboratories across the world and so the physics behind the Sun's proposed energy-generating mechanism was becoming well understood. Although no one really doubted Eddington's model, the detection of solar neutrinos would be an exacting test of its validity. In effect, it would be like sticking a detector into the heart of the Sun. If nothing else, it would make the ghost of James Jeans, who had died in 1946, rest a little easier. Eddington, too, was in his grave by this point, having passed away in 1944.

The two physicists who took up the challenge were Raymond Davis, Jr, from the Brookhaven National Laboratory, New York, and John Bahcall, from the California Institute of Technology. Bahcall made the calculations about neutrinos and detectors, while Davis set about the task of actually building the instrumentation.

To shield it from the false signals generated by natural radiation, they placed the detector as far underground as possible. They found a perfect location in the Black Hills of Dakota, at the Homestake Mine. This was America's deepest mine. The excavation had begun in 1876 during the Black Hill's Gold Rush, and when Davis and Bahcall were working there in the 1960s, the precious metal was still being hauled from ground.

The detector they constructed was a cylindrical tank

20 feet in diameter by 48 feet long. It was placed almost a kilometre and a half underground and contained 100,000 gallons of tetrachloroethylene. This substance was none other than dry-cleaning fluid but contained a specific form of chlorine that could be transformed into radioactive argon when struck by a neutrino.

Theory suggested that a tank that size might catch one neutrino per day out of the 10 million billion that the Sun was expected to be kicking out. So they would allow the argon to build up for a month or two, and then collect it from the tank by shooting jets of helium gas through the fluid. They would ship it back to Brookhaven and measure the quantity from its radioactivity.

The results were monumental. The experiment had indeed detected neutrinos. Although there were some details to sort out, overall there was no doubt any more that the Sun was a vast nuclear reactor, making power by fusing hydrogen into helium. And the path that had led to this achievement had all begun with the nib of Pauli's pen scratching mathematics on to a sheet of paper. It was a stupendous show of force for the particle physicists, and they weren't finished yet. A similar story led to the discovery of antimatter.

In 1928, British physicist Paul Dirac was applying Einstein's special theory of relativity to the quantum mechanics of hydrogen atoms. In the tangle of equations he created, he saw that a positively charged electron seemed entirely possible. Yet such a particle had never been observed in the laboratory.

With the theory published, it took just a few years for someone to find it. American Carl Anderson discovered the positron, as he called it, on 2 August 1932. Since then, antimatter has become a familiar part of physics and science fiction. It's most well-known quality is that it annihilates completely into energy when it meets its matter counterpart. In the case of a positron, whenever it collides with an electron, the pair transform into rays of light.

It is the most efficient conversion of matter into energy that it is possible to have, which explains why *Star Trek*'s Scotty powers the USS *Enterprise* with matter–antimatter collisions. Back in the realms of reality, with neutrinos and antimatter under their belts, the particle physicists were feeling very confident and ready to take on their biggest challenge yet.

To complete the quantum revolution, they needed a theory to explain gravity as an exchange of particles rather than as the warping of spacetime in Einstein's general relativity. In the 1970s, they found something that appeared to fit the bill.

They were busy developing a quantum theory of the strong nuclear force, and were weighing two rival approaches. The first was to suggest that protons and neutrons were made of constituent particles dubbed quarks that communicated the strong force to each other via particles called gluons. The second approach was to suggest that the particles of nature were not minuscule billiard balls but knotty subatomic strings. Both approaches gave some additional mathematical freedom to the theorists.

The argument was settled by a succession of particle accelerator experiments during the late 1960s and 1970s in which quarks were finally observed. For a while it looked as if string theory was redundant, but then physicists began to notice that buried in its mathematics was a description for a very interesting-looking particle. Although hypothetical, it looked like one that could carry the force of gravity. They began referring to it as the graviton.

Physicists had calculated the necessary properties of a gravity-carrying particle based on the behaviour of gravity in general relativity and now here it was popping out unexpectedly from string theory. Perhaps it was the holy grail: a quantum theory of gravity. If so, it would make astronomers very happy.

As we saw in Chapter 5, general relativity cannot tell astronomers what lies at the centre of a black hole, where gravity is extremely strong and changeable over minuscule distances. Such minute scales are exactly where quantum theories work best. And with each new quantum success, the particle physicists had predicted and then found previously undiscovered particles of nature. Could the same be true for gravity?

By the turn of the twenty-first century, the physicists had turned almost wholesale towards string theory. This was because further work was giving hints that string theory might not be just an explanation of quantum gravity after all but an umbrella theory for all the fundamental forces. According to its proponents, it has the potential to explain all particles and forces of nature in a single overarching mathematical framework, often referred to as a theory of everything.

The catch is that to transform string theory into this theory of everything requires an additional hypothetical idea to be true. Called supersymmetry, it postulates that for every known type of particle, there is a counterpart. If true, it means that there are vast swathes of undiscovered particles drifting through the Universe. One particularly sluggish supersymmetric particle is called the neutralino and would be exactly what the astronomers need for dark matter.

Weighing in about 1000 times heavier than a proton, the neutralino is a WIMP (Weakly Interacting Massive Particle) and interacts through gravity and the weak nuclear force of nature alone. This latter interaction is good news because it means that neutralinos will be detectable – if they truly exist.

By 2008, detectors had reached the sensitivity that the theoreticians were saying were needed to detect neutralinos. It was just a question of wiring them up and sitting back and waiting for something to happen.

When something did happen, it wasn't what anyone was expecting.

I was in Munich. It was the beginning of September 2011 and the weather was still hot. I was running on adrenaline because I had just an hour to file my story. The trouble was, I still didn't really know what the story was.

I'd been despatched by *New Scientist* to attend the Topics in Astroparticle and Underground Physics conference because the organizers had called a press conference to announce the results of a dark matter detection experiment called CRESST II. The rumour was that CRESST II had seen evidence of dark matter but the announcement would take place on Tuesday afternoon, just two hours before the magazine went to print. Half a page had been saved. I had to deliver.

Luckily for me, the technical paper was released on an academic website that morning, so I began reading. There was also a detailed presentation by one of the team to the conference before lunch. But the all-important interview, when I could check the subtleties and place it all in context, wasn't going to happen until the very last minute.

Working as fast as I dared, I pieced it all together: the bottom line was that between June 2009 and April 2010, the CRESST II detector had pinged twenty times with a signal they couldn't explain with known particles. But if they were dark matter WIMPs, then they were ten times lighter than the theoreticians were expecting.

CRESST II is located underneath the Gran Sasso mountain in Italy and it was not the first experiment to see a hint of dark matter. That accolade went to sister experiment DAMA in 2008, also sitting under Gran Sasso. Another experiment, CoGeNT, located in a mine in Soudan, Minnesota, announced possible detections shortly afterwards. Just next door in both cases were other WIMP detectors: XENON100 inside Gran Sasso, and CDMS in Soudan. Puzzlingly, neither of these experiments had seen anything at the time.

Even among the three experiments that did have candidate signals, there was no real agreement. Although all three were suggesting WIMPS about ten times lighter than expected, there was overlap on the actual details of the masses and reaction rates. It was a mess of a situation. None of the experiments were seeing anything that was truly convincing, and feelings were running high.

Having dispatched the news piece,* I continued investigating, asking questions and generally trying to get to the bottom of what was going. I swiftly found myself in the swamp of high-pressure science. There can be no doubt that the person who leads the team that finds dark matter will receive a Nobel Prize. And that makes it all extraordinarily competitive.

Academic conferences are usually civilized affairs even

* www.newscientist.com/article/dn20875-third-experiment-sees-hints-of-dark-matter.html#.VVm5zmDV24s

when discussion turns to contentious subjects. At worst, there is a wall of passive aggressive silence. In Munich, I discovered that dark matter hunters are made of bolder stuff.

After the CRESST II presentation that morning, the other dark matter experiment teams had presented their latest results. One member of the CDMS team became particularly vexed and hectored the representative from CoGeNT. The DAMA team tried to say as little as possible. Even when I interviewed them for a more in-depth piece about the situation,* two Italian scientists from the higher echelons of the collaboration just looked at me as if they were on trial and answered my questions with as little information as possible. It was as if they were afraid of perjuring themselves.

I did have some sympathy. No matter how much I wanted them to open up, they wanted to highlight the fact that their detector was seeing 'something' but they didn't know whether it was a neutralino from supersymmetry or not. This highlighted the difference between observational science and theoretical science. Observations can show us that something is taking place but making sense of what is happening is the role of theory, and every time theory gets involved, it lets in the spectre of interpretation. This is unavoidable; it is the cost of doing science. How much easier it would be if particles arrived in experiments wearing T-shirts emblazoned with their names. Instead, scientists have to back-calculate their properties using the detector's measurements and a theory that they assume to be true, such as supersymmetry.

In the case of the DAMA team, as with everybody else, when they turned their signal into a neutralino, they got a

* www.newscientist.com/article/mg21328461.900-dark-matter-mysteries-a-true-game-of-shadows.html#.VVm6W2DV24s

mass that was about ten times too light. So, they insisted that they were not claiming detection of a neutralino, just a detection of something.

This was driven home to me when I was chatting later with Rafael Lang, a member of the XENON100 team at Purdue University in Indiana. He had a beautiful way of putting it. He told me that the CRESST II results and those of the two other experiments might be the first glimpses of something completely unexpected. Instead of dark matter as we think of it, he said that we could be seeing just the highest peaks of some extraordinary new subatomic physics landscape. Or it might be just noise and we are simply learning more about the behaviour of the detectors. A final possibility is that supersymmetry is incorrect.

As well as the lack of convincing neutralino detections from the dark matter experiments, CERN's Large Hadron Collider (LHC) has failed to produce them. With a circumference of 27 kilometres, the underground European particle accelerator should have made them by the bucketload in its collisions. But it hasn't. Indeed, there is not a single piece of evidence in favour of supersymmetry that has emerged from the LHC.

This is not to say that the LHC is a waste of 6 billion euros. Far from it; its discovery of the Higgs particle is exceptionally good science and proof once more that particles can be found on paper first. Every time the LHC fires its beams, it is showing us how the Universe works on the smallest, most energetic scales that we can yet investigate. It is revealing more and more about nature, but if supersymmetry does bow out, the astronomers' best candidate for dark matter – the neutralino – goes with it. Dark matter will be almost back to square one. What then? Do the theorists simply dream up new ideas and off we go again? Although this may be the path of least resistance, the very concept of dark matter is facing other problems.

In the preface to Arthur Eddington's revolutionary book *The Internal Constitution of the Stars*, he states: 'It would be hard to say whether the star or the electron is the hero of our epic.' He was referring to the way in which a whole star's external appearance is based upon its generation of energy through the collision of atoms in its heart. Today, we interpret a galaxy's appearance as being based upon the behaviour of its dark matter.

Because dark matter outweighs ordinary matter, it is expected to sculpt the galaxies into shape, determining their number, as well as their sizes and distribution. Computer models of galaxy formation can then be tweaked until they fit what we see – and that's when the problems start. Fiddling with the models until they give the correct number of large galaxies always offers too many smaller galaxies. Conversely, if you fit the number of small galaxies, then you do not get enough large galaxies. This is a conundrum that has been recognized for decades.

The problem is that neutralinos are cold dark matter, which means that they are heavy particles and slow moving. This combination makes them highly susceptible to clumping, which leads to tens or hundreds of times more dwarf galaxies than anyone can find. Initially, astronomers thought that they just needed to build better telescopes, but the better the observations at showing the distribution of large and small galaxies across the cosmos, the bigger the problems have become. It is an inescapable fact that there are simply not enough dwarf galaxies in the cosmos.

Modern models also predict that clumps of cold dark matter about the size of the Solar System should be floating through our galaxy. They should betray themselves by distorting the light of more distant stars, yet no one has seen evidence for this.

All in all, dark matter cannot be as cold as theorists were predicting.

At the opposite end of the spectrum is hot dark matter. These would be particles like the neutrino – extremely lightweight and moving at close to the speed of light. The trouble here is that simulations show that hot dark matter would wash out much of the structure we see around us today: there simply wouldn't be so many galaxies and clusters of galaxies. So, again, progress is stymied.

The obvious way to go is to postulate a middleweight particle that moves at the celestial equivalent of a jog. This is referred to as (you've guessed it) warm dark matter. Such particles would have no more than one-thousandth the mass of WIMPs so they would resist clumping on the smaller scales, explaining the dearth of dwarf galaxies.

As to what such warm dark matter might actually be, some particle physicists are exploring the idea that there is a new type of neutrino, a heavier cousin to those known, which interacts through gravity alone. It has been dubbed a sterile neutrino.

It is hard to escape a sense of déjà vu these days when reading about new candidates for dark matter. There is absolutely no overwhelming evidence that there has to be dark matter, yet astronomical results are routinely interpreted within its framework. We badly need a reality check – a way of knowing whether dark matter is the avenue to progress or a cul-de-sac where we have been trapped for almost a century. As luck would have it – and it really is luck – that's where the European Space Agency's improbably named LISA Pathfinder mission comes in.

This mission was never intended to contribute to the search for dark matter. Instead, it is designed to test the technology needed to detect gravitational waves. These are ephemeral ripples in the fabric of the Universe sparked by the collision of stars, the formation of black holes, and the great violence of the Big Bang itself. They should be all around us, rather like ripples on a pond.

They are predicted by general relativity to be about one-thousandth the width of the smallest atomic nucleus. That makes them fiendishly difficult to detect, and so before the full Laser Interferometer Space Antenna (LISA) mission is attempted sometime around 2028–36, ESA wanted to test the detection technology.

At the heart of the Pathfinder mission are two identical blocks of metal, each made from 2 kilograms of gold and platinum. Yet these metal hearts do anything but beat. When the mission launches, sometime after summer 2015, they will be the stillest things in the Solar System.

Clamped tightly for launch so that the vibration and g-forces do not damage them, once the spacecraft is in orbit the cubes will be gently released to float freely inside. They will be separated by a distance of about 35 centimetres and an onboard laser will monitor them for changes in their relative position. It will detect movement between the masses as small as a picometre, or a thousandth of a millionth of a millimetre (10^{-12} metres). The movement between the test masses will happen because despite being so close together they will feel slightly different forces of gravity. This gossamer difference will be measured by the onboard laser system.

If all goes well, the next time such masses fly, instead of being just 35 centimetres apart, they will be on three different spacecraft, 5 million kilometres apart. At this distance they will be able to detect the slight jostling of the spacecraft as a gravitational wave ripples by.

The twist is that a group of engineers and scientists realized that LISA Pathfinder's sensitivity will allow it to do much more than just test the technology. It could show us whether we truly need dark matter, or whether we need to modify Newton's theory of gravity. To that end, they propose extending the mission to conduct the greatest gravitational experiment since Arthur Eddington's eclipse expedition of 1919.

Recall Milgrom's MOND idea from Chapter 6. The Israeli physicist found that by tinkering with the mathematics of Newtonian gravity, he could reproduce galaxy rotation curves without the need for dark matter. It required gravity to pull a little harder than expected below a certain critical acceleration, but it is impossible to create these heartbreakingly small accelerations to test the idea in the lab.

In 2006, Jacob Bekenstein of the Hebrew University of Jerusalem and João Magueijo of Imperial College, London, realized that there were points in the Solar System where such accelerations would be expected – and some of those locations would be relatively close to Earth. Their calculations revealed 'saddle points' where the gravity of all the planets, the moons and the Sun would cancel out. If LISA Pathfinder can be made to pass through or close to one of these, then the laser system could test Newton's inverse square law of gravity down into the regime where Milgrom said MOND would take over.

I listened to the details of this idea on a dull grey afternoon, several storeys up in East London, at a seminar organized by Queen Mary University of London and given by Magueijo's PhD student, Ali Mozaffari of Imperial College. I watched in fascination as he began speaking. At mention of MOND, many attendees seemed to switch off. One of the academics pointedly dropped his pen and crossed his arms. But as Mozaffari put his case for a simple, clean test to determine whether there is, as he called it, 'gravity plus', I noticed a change. One jaw even dropped at the accuracy with which LISA Pathfinder could take this measurement.

Mozaffari explained that the closer the spacecraft gets to the saddle point, the greater the precision of the test. If the instrument works as expected and the spacecraft can approach to within 50 kilometres, a MOND-like modification of gravity would rise twenty-five times above the detection threshold. Even a catastrophic miss by about 400 kilometres could still yield a strong enough signal to claim discovery.

A positive signal would be the essential catalyst to dumping ideas about dark matter and focusing back on the fundamental physics of gravity, to develop a new theory and usher in a new era every bit as revolutionary as that heralded by Newton and Einstein.

On the flip side, getting close to the saddle point and seeing nothing unexpected would allow the tightest of constraints to be placed upon modified gravity theories, perhaps even ruling them out completely. In this case, we could say categorically that dark matter is needed. We will have to redouble our efforts to work out the nature of this mysterious matter. Again, it will cause a revolution in science.

Whatever the solution to this problem, it will be a massive leap forward from the unknown Universe we inhabit today. The trouble is that, even before this mystery is solved, astronomers have identified an even greater one. It all started with a suggestion so outlandish that Einstein thought the person making it must be mad.

8

The Day Without Yesterday

Einstein was in the Belgian capital city of Brussels when a man accosted him in the leaf-strewn pathways of Leopold Park. It was October and autumn was in full swing. Einstein was in town for the prestigious Solvay Conference, inaugurated by the Belgian industrialist Ernest Solvay. Designed for the 'finest' minds, it was an invitation-only event. Einstein was on the guest list; the stranger was not.

It is unclear when the man had arrived in Brussels or how long he had waited outside the conference venue, but when he spied Einstein, he approached immediately and set about explaining his ideas face-to-face. He was wearing the garb of a Roman Catholic priest and claimed to have discovered something amazing in Einstein's equations concerning the behaviour of the Universe. If true, it would bring accepted wisdom crashing to the ground.

Georges Lemaître was a volunteer in the Belgian army when he first read about Einstein's theory. The First World War had interrupted his ambition to become an engineer and set him on a different path. He realized that his own mathematical ability far exceeded that needed for engineering, and so taught himself the mathematics required to understand general relativity. He received his degree in mathematics in 1920 and while that took care of Lemaître intellectually, there was also a spirituality in him that needed satisfaction.

He enrolled in a Jesuit seminary and was ordained in 1923. He then gained a scholarship to study with Arthur Eddington in England.

Lemaître's pedigree could not have been better, and his insight concerned the behaviour of the spacetime continuum, Einstein's conceptual fabric of the Universe. Lemaître had seen that it had to be either expanding or contracting but could not be static.

Einstein had also seen this, and found it troubling because the accepted wisdom was that the Universe was unchanging. As a result, he had introduced a new mathematical term to provide an antigravity force that he believed would hold the Universe in place. It was unclear what this term represented in reality, but the observations seemed to demand it: he called it the cosmological constant.

Lemaître saw deeper than that. The cosmological constant was not sufficient to keep the Universe static. It was the equivalent of trying to balance a pencil on its point. While it is possible to get the balance momentarily correct, any small disturbance would break the equilibrium and start it expanding again.

Following his work with Eddington in England, Lemaître had spent time across the Atlantic with American astronomers. First, he studied with Harlow Shapley on the east coast and then with Edwin Hubble on the west, looking for a way to observe the expansion. He found it in the phenomenon of redshift.

Redshift had been discovered in 1912 by Vesto Slipher, an astronomer working at the Lowell Observatory in Flagstaff, Arizona. Whereas the Harvard astronomers were busy taking spectra of stars, Slipher had painstaking applied the technique to the much fainter spiral galaxies. The results showed that most of the spirals appeared to be moving away extremely fast – much faster than any star. This movement stretched the starlight, and since red light has a longer wavelength than blue light, the phenomenon became known as redshift.

Armed with the various redshift observations that had been published in the journals, Lemaître analysed them in the context of general relativity and found that he could derive the rate at which the Universe was expanding. He predicted that any subsequently observed galaxy would fit into the same scheme.

He published the results in the *Annals of the Scientific Society of Brussels* in 1927. Written in French, the paper was entitled 'Un Univers homogène de masse constante et de rayon croissant rendant compte de la vitesse radiale des nébuleuses extragalactiques' (A homogeneous Universe of constant mass and growing radius accounting for the radial velocity of extragalactic nebulae).*

The result – no less than a prediction of the expanding Universe – was completely ignored. It seems unfathomable now that this could have happened. It was mind-blowing science of the purest form. Perhaps it was because the paper was written in French and published in Belgium rather than in one of the more prestigious and more widely read German or English journals. Lemaître himself seemed to anticipate this because he sent a copy to Eddington in England for translation into English and inclusion in the Royal Astronomical Society's journal *Monthly Notices*.

* http://articles.adsabs.harvard.edu/cgi-bin/nph-iarticle_query?1927ASSB...47...49L&data_type=PDF_HIGH&whole_paper=YES&type=PRINTER&filetype=.pdf

There can be little doubt that appearing in English under the auspices of Arthur Eddington would have garnered the work more attention, but it languished on the Cambridge astronomer's desk untouched, perhaps even unread. Lemaître also sent a copy to Albert Einstein, who failed to respond, and this is what made Lemaître decide to pay the great man a visit in Leopold Park.

The Einstein that Lemaître encountered that day was not the figure of triumph from a decade before. Now he was a besieged intellect, and becoming intransigent as he fought for the mastery he once had. The root of his problem was quantum mechanics. Einstein himself had helped pioneer this topic in 1905 by showing that light could be thought of as photons. But Einstein was unsatisfied because his equations did not allow him to predict the direction in which atoms would emit light. It seemed completely random. The equations also did not allow the calculation of how long an atom would hold some energy before emitting it as a photon. That, too, seemed random and so was unacceptable to Einstein.

Chance was something that Einstein considered impossible in nature. His belief was that the Universe ran to order and rule, and in that case, everything should be calculable. If chance plays a role in nature then he knew it weakened the role of science, because at the heart of the discipline's success was the ability to predict future events. These could be as esoteric as being able to calculate the deflection of starlight around the Sun or as practical as knowing whether a house would fall down if it were built in the wrong way. But they were both testable, and so provided the route to falsifying and therefore discarding wrong hypotheses.

So he thought that there must be a deeper theory that

would restore determinism and therefore testability. Other scientists such as Niels Bohr and Werner Heisenberg were not so encumbered. They were interested only in what they could measure, and if chance appeared to be hard-wired into the Universe, then so be it.

Einstein's frustration boiled over at Solvay that year, and is the first time he is said to have made the statement that God does not play dice. To which Bohr is said to have rejoined, 'Einstein, stop telling God what to do.'

With control of quantum mechanics slipping from his grasp, Einstein was in no mood to contemplate new ideas in relativity, which he also felt was being appropriated by others. So when Lemaître turned up, he saw not a colleague but another would-be usurper. The Roman Catholic collar the priest wore would not have improved Einstein's disposition either.

The ensuing conversation was strained. Einstein admitted to receiving the paper but dismissed it on several grounds. First, he said it wasn't original. A similar paper extolling the possibility of an expanding Universe had been written in 1922 by a Russian mathematician called Friedmann. Second, although the mathematics was sound, Einstein said the physics was appalling. For this we should read that it didn't satisfy his idea of what the Universe should be doing.

Lemaître argued his case using the redshift measurements as evidence. Einstein had not heard of these but was still reluctant to give ground. Although the two spent the afternoon together with one of Einstein's former students, no real headway was made. Meanwhile, in America, Edwin Hubble was hard at work.

• • •

The anglophile astronomer had already shown that galaxies were whole collections of stars at large distances (see chapter 6). Now he was extending his measurements to even further galaxies. Right back at the beginning of his career, in the summer of 1914, Hubble had attended the American Astronomical Society's meeting in Evanston, Illinois. There he heard Slipher describe redshift. The subject must surely have come up again when Lemaître visited him in 1925. Although the details of that visit are lost to history, what is certain is that soon afterwards Hubble and observatory assistant Milton Humason began accumulating as many redshift measurements as they could.

In 1929, Hubble published results that ignited science in the way Lemaître's paper of two years earlier should have done. He showed that the further away a galaxy was, the greater was its redshift. It was definitive proof that the Universe was expanding, exactly as Lemaître had predicted, yet Hubble did not include a single reference to him or his work. All he did was obliquely mention that the results were applicable to general relativity's description of the Universe.

It took years for Einstein to comment publicly on Hubble's findings, and when he did, he claimed that the cosmological constant he had introduced to hold the Universe steady was his 'greatest mistake'. He removed it from his equations and died in 1955 believing it was superfluous.

Hubble's results roused Eddington, who hurriedly organized for Lemaître's paper to be translated and printed. However, Lemaître asked for the crucial prediction of the rate of the Universe's expansion to be dropped from the article. Maybe he thought that it would look out of date to include it now. Whatever the reason, it made it look as if the work was more a response to Hubble's discovery than a presaging document.*

So Hubble went down in history as the person who discovered the expanding Universe, yet in reality it was an honour that should be shared by Lemaître, and maybe others too. Back in 1924, Swedish astronomer Knut Lundmark estimated the expansion rate of the Universe to within 1 per cent of its modern accepted value. This was three years before Lemaître's paper and five years before Hubble. Both Hubble and Lemaître got their estimates wrong by almost a factor of ten.

But neither Lundmark nor Hubble saw the implications of the expanding Universe in the way Lemaître did. If the Universe was expanding now, then everything had to have been closer together in the past. The question was how close? Lemaître toiled at solving the problem, delving into the mathematics to look for a time in the past beyond which the Universe could not shrink any smaller.

He called this endpoint the primeval atom and took his inspiration from the discoveries of radioactivity. He reasoned that all matter may have once been compressed into a single whole 'atom', which then spontaneously split apart and the resultant debris was carried to the further reaches of space by the expansion of the Universe.

When he looked at the equations, he found that there was nothing to stop time running all the way back to zero. The whole Universe would have been a singularity, a point of infinite density and zero volume. While initially calling this the bottom of space and time, he then thought of something

* Something of a conspiracy theory sprang up around the deletion of the crucial prediction. Mario Livio of the Space Telescope Science Institute unravelled the circumstances in a paper that can be found at http://hubblesite.org/pubinfo/pdf/2011/36/pdf.pdf

better and referred to it as 'the day without yesterday', the moment when time and space began.

The year was 1931, and despite the redshift measurements that proved the expanding Universe, still no one was inclined to entertain Lemaître's idea of a beginning. Scientifically, there was no observational evidence that the Universe began in such a dense state, and personally, many were sceptical of his motivations. They thought his religious beliefs were colouring his judgement and that this was an attempt to foist a biblical creation onto science.

One person who did take Lemaître seriously was George Gamow, a Ukrainian physicist who defected from the Soviet Union in 1931, the same year that Lemaître published his ideas about the beginning of the Universe.

Repeatedly denied travel permits to conferences by the authorities, Gamow tried twice to kayak his way to freedom. Both times he was stymied by the weather. Then out of the blue, he was granted permission to attend a Solvay Conference, again in Brussels. Arguing the need for his wife, who was also a physicist, to accompany him, they both set off and never returned.

By 1948, Gamow was a naturalized American citizen, ensconced at George Washington University and fascinated by the consequences of Lemaître's day without yesterday. He reasoned that if everything had been closer together in the past, then, just as an expanding gas cools down, the earlier phases of the Universe must have been hotter than today. Together with his student Ralph Alpher, he showed that there would have been a time when the entire Universe was at a similar temperature to that which Eddington had found for the Sun's interior.

If so, the whole Universe would have briefly become a nuclear reactor, fusing hydrogen into helium. Gamow and Alpher calculated the consequences, revealing that about a quarter of the hydrogen should have been turned into helium. It matched spectacularly well with observations. The abundance of helium in the Universe today is measured to be about 23 per cent by mass, with hydrogen at 75 per cent and all of the other elements just 2 per cent combined.

It was dramatic circumstantial evidence that the Universe had a beginning and had evolved into what we see today, but it wasn't proof. For that they needed a testable prediction. Alpher and a colleague, Robert Herman, took up the challenge.

Gamow and Alpher had mentioned that the leftover radiation from the cosmic helium-building maelstrom should still fill the Universe. Alpher and Herman calculated the strength of this radiation today, and predicted that it would give the Universe a temperature of about 5 °C above absolute zero. Observe this, and it would prove Lemaître and Gamow's vision of the early Universe.

There the matter rested until the mid-1960s, when radio engineers Arno Penzias and Robert Wilson accidentally discovered the all-pervading hiss of microwaves bathing the Universe. Its temperature was 2.7 °C above absolute zero, just a whisker off Alpher and Herman's prediction. This was the cosmic microwave background radiation that we discussed in the Introduction.

Its presence meant that there could no longer be any doubt that the Universe had begun in a hot, dense state. News of his vindication reached Lemaître as he lay on his deathbed in Leuven, aged seventy-one. In time, the day without yesterday became known as the Big Bang.

But how hot and how dense could it have been? Lemaître's ideas placed no barrier on these quantities. The idea of the Big Bang as an infinitely dense, zero-volume instant in time, out of which the observable Universe sprang, seemed unavoidable. In many ways it resembles a black hole. Both are mathematical singularities which require a quantum theory of gravity to understand. Until we have that, anything a cosmologist says about the nature of the Big Bang must be taken with a large pinch of salt. Or so we thought.

In March 2014, a small group of observers announced the detection of a signal that could only have been generated a minuscule fraction of a second after the Big Bang. Our confidence in what happened during the birth of the Universe was poised to break into a whole new regime of certainty.

It began for me with a telephone call, and a simple question from the *Guardian*'s assistant national news editor, James Randerson. 'What do you know about gravitational waves?' he asked.

Quite a lot, as it happened. Several months earlier I had written about the LISA Pathfinder mission from the last chapter for *New Scientist*.

'We're hearing rumours that they've been discovered in the microwave background radiation,' said Randerson.

That stopped me dead. These were not the gravitational waves from exploding or colliding stars; James was talking about primordial gravitational waves generated from a moment as near to the Big Bang as we are ever likely to see. Frankly, it was jaw-dropping. It would prove that Einstein's spacetime continuum adhered to quantum mechanics because primordial gravitational waves were hypothesized to come about through the action of one of quantum mechanic's foundation stones.

Called Heisenberg's uncertainty principle, we encountered it first in Chapter 3, when it saved Eddington's model for the Sun's interior by lowering the temperature at which nuclear reactions could take place. It ties together measurable quantities such as position and momentum, time and energy, and describes how the more precisely one is measured, the more uncertain the other becomes.

The question we have to answer today is whether it applies to spacetime as well as to particles. If it does, then instead of the smooth continuum that Einstein imagined, spacetime would be made of discrete particles, rather like a beach is made of sand grains. The properties of these grains of spacetime will fluctuate within the boundaries set by the uncertainty principle and this will create the primordial gravitational waves. Usually these would be far too small for us to detect, but cosmologists had been toying with an amplifying mechanism that came into play just after the Big Bang.

They call the hypothesis 'inflation' and have been collectively developing it for decades. It is an ad hoc piece of mathematics, bolted on to our ideas of the Big Bang to help solve some tricky inconsistencies related to the cosmic microwave background radiation.

In the late 1960s, gravitational physicist Charles Misner pointed out that while the microwave background radiation was coming from all parts of the sky, just as you would expect from the Big Bang, it was completely uniform and that was a big problem.

Everything we know about physics tells us that the temperature of the radiation on one side of the Universe should be different from that on the other. This is because one side of the Universe cannot yet know of the other's existence. Separated by more than 90 billion light years, the opposite sides have not had time in the Universe's lifetime of 13.8 billion years to exchange energy and equalize temperature. Yet, the entire Universe displays the same temperature, no matter where astronomers look.

If one dramatic cosmological problem wasn't enough, another one swiftly followed, care of Robert Dicke in 1969. This one is called the flatness problem.

Einstein's equations show that matter and energy curve the spacetime continuum. This means that the Universe should have an overall curvature based upon the average density of matter and energy it contains. The puzzle is that, as far as anyone can tell, the density is poised on a highly unlikely value: the critical density to give the Universe no overall curvature. It represents a kind of perfectly balanced cosmos.

A third great mystery was recognized by the particle physicists. If their theories are right, the Universe should be filled with particles called magnetic monopoles. But to date, not a single one of these has been found.

Late one night in 1979, young researcher Alan Guth was working overtime. He was toying with the physics of the early Universe, wondering if it could experience a phase transition. These are familiar to us all; they occur when matter changes from one state to another – solid to liquid, for example. Ice cubes melting are a phase transition. Guth was looking at the way in which the spacetime continuum may experience its own version of a phase transition and realized that such a sudden release of energy would drive the spacetime continuum into a sudden exponential jolt of expansion. Guth realized that it could solve the horizon, the flatness, and the monopole problem. On the page of the notebook where he was toying with the ideas, he wrote 'SPECTACULAR REALIZATION' and drew a double box around the words.*

As so often happens in science, Guth wasn't the only one to be thinking along these lines. In Russia, Alexei Starobinsky was working at the Laudau Institute for Theoretical Physics in Moscow. He, too, generated a mathematical framework in which inflation could have happened. Indeed, while the details of Guth's inflationary idea have long been superseded, Starobinsky's remains one of the leading candidates.

Inflation's basic idea is that the Universe was born with a vast reserve of energy locked into the spacetime continuum. Something jolted it to start shedding this energy in a cascade that spread across the infant Universe. This drove inflation, which doubled the size of the Universe about eighty times in just 10^{-36} seconds. But don't go thinking that the Universe was big back then. Back-calculating from the size of everything we see around us today, it appears that our observable Universe all grew from a patch just 10^{-28} metres across, a billion billion times smaller than a hydrogen atom. During inflation, that infinitesimal mote became no more than a centimetre across.

By suddenly inflating in this way, energy was smeared out across the cosmos, equalizing the temperature everywhere; spacetime was stretched so much that any initial curvature became imperceptible, and the monopole particles were flung to the very farthest reaches of space. In effect, this turns the

* Dennis Overbye provides a good narrative of this: www.nytimes.com/2014/03/18/science/space/detection-of-waves-in-space-buttresses-landmark-theory-of-big-bang.html?_r=0

Universe into a blank canvas, a featureless flat sheet of spacetime in which gravity can start pulling matter together to form stars and galaxies. The only details that survive are the quantum jittering of spacetime, which are amplified by the inflation into large-scale gravitational waves that disturb the cosmic microwave background in an observable way.

As an idea, inflation certainly ticked all the boxes, but there was very little to link it to testable reality or to the rest of physics. For example, the nature of the energy field that powered inflation remains entirely unclear even now. Despite these shortcomings, it has gained acceptance as the way the early Universe probably behaved. Echoing Eddington's stand over the criticism of his model of stellar energy generation, inflationary cosmologists say that although the hypothesis may not provide all the answers, it's the best explanation we have on offer.

The belief is that continued toil will bring it to completion, but for the moment inflation is conviction cosmology; working on it is an act of faith as much as science. Its one salvation is that it does have a prediction to make: it would imprint primordial gravitational waves into the cosmic microwave background. Find these waves and, despite all its shortcomings, inflation has to be correct. And this was what James had phoned me to say the cosmologists were preparing to announce.

The team at the centre of the hoo-ha were a mostly American consortium of scientists who worked on an experiment called BICEP2. This stood for the Background Imaging of Cosmic Extragalactic Polarization and was the second instrument they had built. It is located at the South Pole, which has a largely unobstructed view of the deep sky, where the cosmic microwave background is brightest.

When the scientists made their announcement at Harvard on Monday, 17 March 2014, it was exactly as the rumours suggested. Not a single cosmologist I spoke to had anything but praise for the work. It seemed as if a genuine epoch-making breakthrough had been achieved. My story appeared online* first and the newspaper ran it on the front page the day afterwards. Then the problems began.

Late on Monday evening, my phone chimed with an email. It was from John Peacock, a senior cosmologist from University of Edinburgh. The message was a simple one: amid all the jubilation, no one was saying anything about the downside. My mind did a somersault: a downside? No one had mentioned this to me. Peacock explained that the apparent vindication of inflation meant that there is a limit to the knowledge we can ever have of the Big Bang itself.

The more I thought about it, the more I realized the truth. Inflation effectively erases all details of what took place before. Everything is smeared to uniformity by the exponential expansion. It was designed to take *any* initial condition and turn it into the Universe we see around us today. The flip side is that we can never know what happened before inflation. There might not even have been a Big Bang. We will never know. Science cannot tell us. Put like that, the discovery doesn't sound so much like a breakthrough as a brick wall.

Next morning, I spoke to David Wands of the University of Portsmouth. He broadly corroborated this more negative slant, telling me: 'Inflation is a way of avoiding a lot of the difficult problems we encounter when discussing the Big Bang. It means that we can describe a lot of the properties of the

* www.theguardian.com/science/2014/mar/14/primordial-gravitational-wave-discovery-physics-bicep

Universe without having to push back to the Big Bang itself.'

Other cosmologists told me they thought such views were too pessimistic. They said there might be some way of extracting information about the Universe before inflation but they had no idea of how to do that yet. When asked on what grounds they based their optimism, they eventually admitted that it was gut instinct alone, rather than a scientifically reasoned position.

Nor might it just be the Big Bang that is hidden from view. As we have discussed in this chapter already, the primordial gravitational waves prove that spacetime is quantum. The catch is that the era of quantum gravity would now be firmly trapped behind the featureless mask of inflation, ruling out direct cosmological tests. We may be forced to abandon the pursuit of this knowledge altogether.

Others in the cosmological community argued that inflation *was* the Big Bang. MIT cosmologist Max Tegmark took to the Internet to advance this idea, redefining what we should mean when we refer to the Big Bang. With typical aplomb he reasoned that the tiny part of space that became the Universe we see around us today was so minuscule that the amount of mass it must have contained was less than an apple. During inflation, vastly more spacetime came into existence, replete with its inherent energy density, which could then turn into matter and radiation once inflation was done. By any definition, he argued, this sounds like what we thought the Big Bang did.*

Before we get carried away with this idea, however, let us remember Lemaître's concept of the day without yesterday.

* www.preposterousuniverse.com/blog/2014/04/21/guest-post-max-tegmark-on-cosmic-inflation/

It was the moment that time and space began. Tegmark admitted that, even with the BICEP2 advance, we still didn't know what – if anything – came before inflation, but his clear implication was that it doesn't matter: inflation is the de facto Big Bang. This appears to be a tacit acknowledgement that science does have limits, confirming Peacock's point. However unappealing that might seem, the announcement of BICEP2 meant that we simply had to deal with it. We had hit the wall in cosmic knowledge.

Then doubts about the detection itself began to creep in. The gravitational waves in the cosmic microwave background were inferred because of the detection of a property called circular polarization. This is a corkscrewing motion in the microwaves implanted by the passage of the gravitational waves. But there is something else that can also induce circular polarization in the microwaves: dust.

At the beginning of this book, I described the image that the European Space Agency's Planck satellite took of the microwave sky. Before that final image in 2013, the agency had released an earlier version. On that one the blueprint of the Universe was largely obscured by a wispy foreground that looked like the clouds one might see on a summer's day being torn apart by the wind. It was microwaves produced by nearby dust grains in our Galaxy, the take-home message being that microwaves can be produced by both the birth throes of the Universe and the dust motes in our celestial backyard.

To get at the cosmological information, the foreground microwaves have to be subtracted from the image. To do this successfully, Planck took readings at nine different frequencies to allow the scientists to model the behaviour of the dust. BICEP2, on the other hand, had data for just one frequency.

On the face of it, this made the removal of the dust contamination impossible. So how had BICEP2 done it?

The story that emerged was that they had downloaded preliminary dust maps shown by Planck scientists at a conference, and applied these to their data. The revelation caused more than a few open mouths among the cosmological community. It wasn't so much the purloining of the data that caused all the headshaking and muttering; it was the fact that the Planck maps were works in progress. They had been shown at a conference as a promise that the work was ongoing, and that the full maps would be released as soon as possible. Until then, no sound conclusion could be based on the data. Yet there was no uncertainty in the Harvard University email that invited journalists to their press conference: they promised us a major discovery.

Stanford University had uploaded a YouTube video (since removed) of a BICEP2 researcher surprising inflationary theoretician Andrei Linde with a bottle of champagne and the news that BICEP2 had found the 'smoking gun evidence of inflation'. Now, it was all starting to unravel.

In the months that followed, the BICEP2 team found themselves increasingly on the back foot over both their science and their methodology. When the paper describing their results was finally published on 19 June 2014, the dodgy Planck data had been removed and the confidence in their results had dropped significantly. At the very moment the paper was published, I was at University College London, in a lecture being given by Clement Pryke of the University of Minnesota, a co-leader of the BICEP2 team. Celeste Biever, the physical sciences news editor at *New Scientist*, was feeding the details of the newly released paper to my iPhone, and I was trying to assimilate all of that and listen to the lecture.

Pryke had refused my request to interview him before his talk, and appeared to be taken by surprise when I told him in the question and answer session that his paper had just been published. Indeed, he had finished his presentation by telling the audience that the team were still working on the final submission.

All in all, I felt rather sorry for him. He had a beleaguered aura that was painfully apparent when a straight question about his feelings over the result led to the confession: 'Has my confidence gone down? Yes.' Sadly for Pryke and the others, worse was to come.

On 22 September, without fanfare, the European Space Agency's Planck satellite team released their findings on the Internet archive known as arXiv (let the weeping begin for those sensitive to spelling; the X in the name is actually the Greek letter chi). The official word from the Planck consortium of more than 200 astronomers around the world was that the signal seen by BICEP2 could easily have come from dust, not inflation.

For the initially triumphant BICEP2 scientists, it was the story of hubris run aground. It certainly highlighted the high risk of claiming an extraordinary discovery before others have had the chance to sift the data and double-check the methodology.*

But if you think that this is a watershed moment for inflation, think again. It does not seem to have dented many people's belief in the theory. A number of other groups are still building telescopes to look for the primordial gravitational

* Ross Anderson has a great feature about these events at http://aeon.co/magazine/science/has-cosmology-run-into-a-creative-crisis/

waves, and many believe a definitive detection is just around the corner. There is nothing wrong in this *per se* but there is a side to inflation that should worry us all. It is this: during the outpouring of theoretical tinkering, so many different versions of inflation have been found that the idea has no unique, testable characteristic. The gravitational waves are the nearest thing, but even these are not definitive because some versions of inflation do not predict them. So, although seeing gravitational waves would confirm inflation, not seeing them would not rule the idea out. In other words, the hypothesis is impervious to contrary data. Where does this situation leave science? Testable predictions must surely remain the gold standard, because it is the only route to certainty.

Cosmologist Paul Steinhardt thinks the same. He gave an interview to John Horgan for *Scientific American*'s blog pages in which he said, 'Scientific ideas should be simple, explanatory, predictive. The inflationary multiverse as currently understood appears to have none of those properties.'*

We'll discuss the multiverse in the next chapter, but for now let's stick to inflation. What makes Steinhardt's view so powerful is that not only is he Albert Einstein Professor of Science and Director of the Centre for Theoretical Science at Princeton University, New Jersey, he is also one of the founding father of inflationary theory, helping to move the idea on from Guth's original conception.

Falling out of love with his own theory because of its ability to absorb just about any observation, he called it a theory of anything.†

* http://blogs.scientificamerican.com/cross-check/physicist-slams-cosmic-theory-he-helped-conceive/

† http://edge.org/response-detail/25405

The majority of theorists would argue that continued observations will allow them to home in on the right variant of their idea, but that presupposes that the idea is correct in the first place. If there is not a final test to rule the idea in or out, then there can be no absolute proof, only inference and, dare I say it, faith that the model is right.

To break this increasingly vicious circle, we need to look at the problem from a different angle. Perhaps then we can gain a new clue and re-energize cosmology. That new clue may have come in the mid-1990s, when cosmologists were looking for an answer not to how the Universe began, but to how it was going to end.

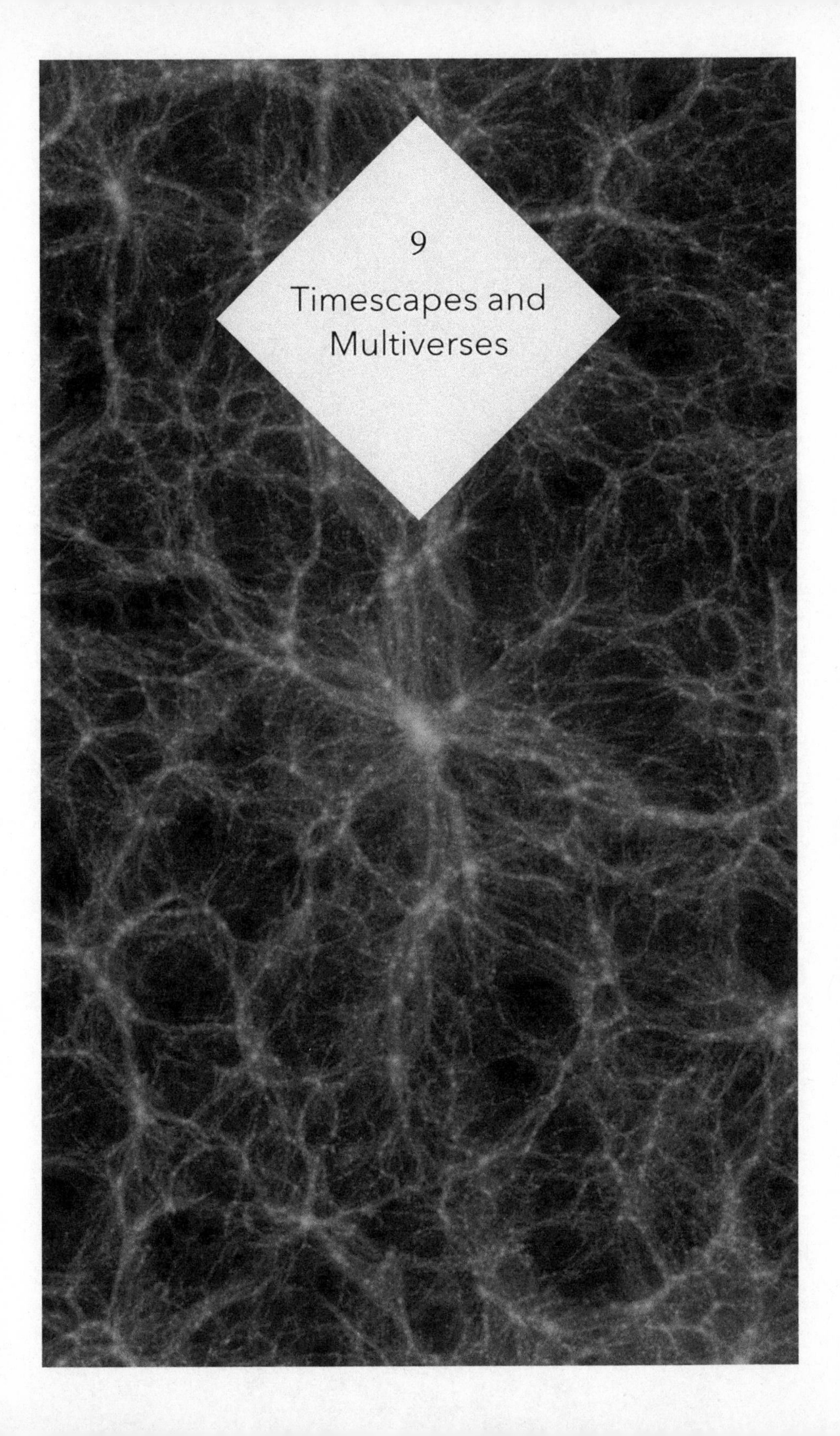

9

Timescapes and Multiverses

'I am constant as the northern star,' makes for great Shakespeare but, as any astronomer will tell you, the bard's words are factually inaccurate. Although the stars appear fixed in position for the duration of our lives, they are moving in their great orbits and slowly creep across the sky, changing the shape of the constellations over the course of tens of thousands of years. Neither are stars immortal.

The energy-generation models pioneered by Eddington can be used to predict their lifespans. Depending upon its mass, a star can live for between 10 million and 100 billion years. The most massive stars live the shortest lives because they use their fuel the fastest. The result is that the stellar population will change with time, as new generations are born and older generations die.

Stars of the future will not be the same as stars of the present. They are likely to be dimmer and longer-lived because they will contain a greater proportion of heavy chemical elements produced by the fusion of hydrogen and helium in the previous generations of stars. At the end of a star's life, these chemicals are returned to space, either in titanic explosions known as supernovae, or in more genteel stellar cataclysms where the dying star gracefully wafts its outer layers into space. The heavier elements have a cooling effect on the interstellar gas clouds, making them more susceptible to gravitational collapse and this results in a larger number of smaller stars, which are naturally dimmer and longer-lived.

With a greater quantity of heavier elements, it should also be easier to build planets. This could mean that the Universe has not yet entered its most habitable phase. In the future there could be more small stars, each with a wealth of planets in orbit around it.

On the larger scale, the most important factor governing the Universe's future is its rate of expansion. As Lemaître showed, this can be calculated from general relativity and is related to the Universe's overall gravitational field. It is possible that the Universe will expand forever, but this is not guaranteed. If there is enough matter, the expansion will slow down and then reverse, speeding everything back together in a cataclysmic 'Big Crunch'. Either way, it is not good news for the Universe.

In the first scenario, known as an open universe, everything will appear more or less business as usual to begin with. Stars will continue to live and die much as they have done for the last 13 billion years, but the Universe itself will change around them.

Space will expand and the galaxy clusters will recede ever farther away from each other. Eventually, the galaxies that make up Herschel's luxuriant garden will be lost from sight because their light will be redshifted away from visible wavelengths into the infrared. As the aeons pass, the redshift will transform the starlight into weak radio signals.

These faint whispers will be the only clue to some future civilization of the existence of other galaxies, but as time continues to pass, the already weak signal will be redshifted even more, rendering it so feeble as to be unobservable. The cosmic microwave background radiation that gave us our first definitive evidence of the Big Bang will also be lost in the same manner. Future civilizations may thus lose the ability to perform cosmology because the Universe beyond our immediate cosmic neighbourhood will be unutterably dark and empty.

The fifty or so nearest galaxies in our own cluster will not recede because they are bound to us by gravity, and that means they must be in some kind of orbit. Eventually these epic circulations will result in the merger of the galaxies with each other.

As the galaxies collide, some stars will be flung out into intergalactic space to wander the Universe alone. Abraham Leob, of the Center for Astrophysics at Harvard University, has suggested that these could be seen with extraordinarily powerful telescopes and used in the same way that receding galaxies are used today to tell us about the expansion of space. Others think the faintness of the stars makes this a very long shot indeed, and besides, none of us is likely to be here in 100 billion years to see if it's possible anyway.

Many stars that are not catapulted out will be tipped into the supermassive black hole at the centre of their galaxy, temporarily reigniting the activity around the lurking behemoths. Eventually, the supermassive black hole at the centre of each merging galaxy will collide, releasing a cataract of gamma rays strong enough to sterilize all remaining planets in the galaxies.

After some 100,000 billion years all the cosmic gas will have been either pulled into existing stars or sucked into black holes. One by one this last generation of stars will run the course of their lives and die, becoming black holes or other stellar remnants.

With no gas clouds left to collapse into new stars, stellar activity will begin to come to an end. The lights will be going out across the cosmos, leaving space to contain nothing but isolated collections of dead stars: white dwarfs, neutron stars, and black holes both large and small. These stellar corpses will occasionally collide, releasing a sudden burst of radiation, but other than that, there will be no light shining through the Universe.

As grim as it sounds, this may not be quite the end. Protons are composed of three quark particles, and in some extensions of our current particle physics theories, this makes them liable to decay into other particles. Experiments searching for the signature began in the early 1980s. Rather like the dark matter experiments, these have so far produced no positive results and plugging that into calculations shows that the half-life of a proton must be something enormous like 1.29×10^{34} years. By comparison, the Universe is only about 10^{10} years old now.

If protons do eventually decay, then all the atoms in the Universe will disintegrate, leaving nothing but a sea of subatomic particles. Chemical reactions and nuclear reactions will become impossible because there will no longer be any particles capable of forming an atomic nucleus, let alone a chemical element or molecule. All these particles will be capable of doing is falling into existing black holes, or clumping together to form new ones of their own. But not even the black holes may last forever.

According to work by Stephen Hawking that we will encounter in the final chapter, they may gradually radiate their matter content back into the Universe in the form of more subatomic particles. Such a phenomenon would take place over an inconceivably long time period, perhaps a googol (1 followed by 100 noughts) years. So, the long-term fate of an open Universe is to become a dilute sea of particles and black holes, all at approximately the same low temperature and unable to react with one another. Such a state is known as the heat death of the Universe.

The other option is if the Universe contains enough matter to collapse back in on itself. This is called a closed universe, and if it is to be our fate then at some point in the future the expansion of the Universe will halt and a contraction will begin.

The galaxies will get closer and closer together, heading for the Big Crunch. Instead of a redshift, there will be a blueshift as the galaxies fall back together. This will make it look as though the stars are heating up because their light will be squashed to shorter wavelengths and, as the lady computers of Harvard found, the stars are classified by the colour of their light, which is linked to their surface temperature. The hotter the star, the more blue light it emits. In reality, the stars in this scenario are not getting hotter; they are simply having their light 'blueshifted' by the collapse of the Universe.

A billion years or so before the final supercollision, clusters of galaxies will merge together. Around 100 million years before the end, individual galaxies will begin merging. For the last million years of the Universe's existence, there will be no such thing as an individual galaxy – the entire Universe will be one great ocean of stars.

During the contraction of the Universe, the cosmic microwave background radiation will be squashed; it will transform the microwaves first into infrared and then into visible light, making the whole night sky light up around 100,000 years before the Big Crunch.

Finally, the blueshifted background radiation will become so intense that it will exceed the temperature of the stars themselves. Rather like ice cubes placed in hot water, the stars will dissolve into space and the Universe will resemble the fireball of the Big Bang. Such is the similarity that some cosmologists have speculated about whether the Big Crunch can turn into another Big Bang and start off the whole process of cosmic evolution again.

In 2010, cosmologist Roger Penrose, from the University of Oxford, and a collaborator Vahe Gurzadyan, from the Yerevan Physics Institute, Armenia, proposed a new solution to Einstein's equations of general relativity called conformal cyclic cosmology.* In this idea, our Universe is simply the latest in a cycle of universes that each spring from the ashes of the previous one. They even claimed that there was evidence to support their ideas in the distribution of temperatures in the microwave background radiation.

Although other cosmologists have disputed this claim, what is certain is that there are 'little crunches' happening every day under astronomers' eyes. They occur in exploding stars known as supernovae. In the very centre of these cataclysms, the density of matter can become so great that the spacetime continuum collapses in on itself, and a black hole is formed. By the 1990s, cosmologists were using such supernovae to determine whether the Universe as a whole would suffer the same fate. The methodology is simple. Astronomers wait for a supernova to explode and then train their telescopes that way. But not just any old supernova would do.

Supernovae come in a variety of types. One in particular is highly prized for this work. Known as a type 1a supernova, it is

* arXiv:1011.3706v1 [astro-ph.CO]

a binary star where one star has died to become a white dwarf, a dense ball of mostly oxygen and carbon that was once the nuclear heart of the fully fledged star. It is about the volume of the Earth but contains the mass of the Sun. As its companion star ages, it can spill gas onto the surface of the white dwarf. Once the white dwarf collects enough of this, it will explode. The tipping point is known as the Chandrasekhar limit and is the equivalent of 1.4 times the mass of the Sun.

Although the precise nature of the detonation is still a matter of scientific investigation, the practical upshot is that because all type 1a supernovae explode with the same mass, they each project the same luminosity into space. Their measured brightness is therefore dependent only on their distance, and this makes them excellent mile markers.

To make use of this, astronomers measure the redshift of the supernova's galaxy. As we have seen, the redshift is directly related to the expansion of the Universe. Once they have done this for a number of different galaxy-supernova combinations at different distances, they compare them with their models of how the Universe is expanding.

The thinking was that all the matter in the Universe must be causing the expansion to slow down, but when cosmologists computed a number of different possibilities and compared them with their observations, none fitted. The supernovae were all too dim for their redshifts. The only way to fit the data was if the expansion of the Universe was not decelerating but speeding up. This was a shock indeed. An accelerating expansion went against everything cosmologists thought they knew about the behaviour of the cosmos.

If it had been a single team of astronomers performing this work, the suspicion of an error would have undoubtedly been raised, but the conclusion was reached by two rival teams more or less simultaneously: the High-Z Supernova Search Team and the Supernova Cosmology Project. Both were racing to be certain of their extraordinary findings and ended up releasing their results within a week of each other.

But what did this crazy conclusion mean? Deceleration was based on the notion that matter (in all its possible forms) was the dominant ingredient in the Universe. Acceleration implied that something with properties opposite those of matter was dominant. Cosmologists were quick to postulate that some exotic form of energy could be exerting a force of antigravity. They began referring to it quite confidently as dark energy, while at the same time scratching their heads and wondering what to do for the best.

In 2006, two independent committees of leading cosmologists were convened, one on either side of the Atlantic, to set an agenda for investigating dark energy.

In Europe, John Peacock of the University of Edinburgh, UK, headed a committee for the European Space Agency (ESA) and the European Southern Observatory. It concluded that, of all the conundrums facing cosmologists, dark energy posed the greatest challenge because there is no 'plausible or natural' explanation for it. In other words, nothing in known physics even hinted at such an exotic substance.

In America, Rocky Kolb of Fermilab in Illinois reached similar conclusions. He chaired the Dark Energy Task Force, which reported to the US Department of Energy, NASA, and the National Science Foundation. His report recommended an 'aggressive program to explore dark energy as fully as possible' because it 'challenges our understanding of fundamental physical laws and the nature of the cosmos'.

Even now, after almost a decade's extra effort, there is not much sign of progress. Dark energy remains as mysterious as ever with no obvious origin for it in anything we know about natural science. We are therefore driven to believe that there is unanticipated physics to be discovered. This is often referred to as 'new physics'. Either there is a new type of energy, or a new force of nature, or gravity behaves in a way that goes beyond general relativity. Any one of these would require an overhaul of our theoretical description of nature.

An obvious place to start the search for an explanation of dark energy is the cosmological constant, that we encountered in Chapter 8. This was the exotic form of energy Einstein flirted with because he couldn't take the expansion of the Universe seriously. It does represent a form of antigravity, but to turn it into a candidate for dark energy would involve bumping up its strength to previously unheard of proportions. To account for the supernova observations, a whopping three-quarters of the Universe must be dark energy.

Then there are the results that deepen the mystery further by suggesting that dark energy has not always been so powerful. The data now suggests that dark energy 'switched on' 7 billion years ago. Before that time, the expansion of the Universe was slowing down as expected. After that time, things began to speed up. This is highly peculiar and indicates a type of behaviour beyond the cosmological constant, which, as the name suggests, should stay constant.

So perhaps dark energy is an unanticipated force of nature. This often gets called quintessence as a nod towards antiquity and the classical elements. To the Greeks, the Universe was made of quintessence, which we could never find on Earth. Modern quintessence is thought to be a weak force of nature.

Although weak, across the sweeps of intergalactic space it mounts up into a potent agent. The trouble with this idea is that to accelerate the Universe in the observed way, the quintessence field should also act as an antigravity force between individual celestial objects. But as we have seen, the reverse is true; there appears to be too much gravity – hence astronomers invented dark matter.

Another possible approach is to assume that gravity behaves differently from the way we expect, pulling a little more weakly on the intergalactic scale. But trying to 'turn down' gravity on the cosmic scale appears virtually impossible to do without affecting motion on the small scale, where the planets all move exactly as predicted by Newton and Einstein.

It is all starting to look like a catch-22, but there is a lesson from history that might help here. When things have looked impossible before, the breakthrough has finally come when a particularly brave scientist has thrown away a cherished assumption. Often this assumption has become so entrenched that many think of it as established fact. Remove it, however, and although the calculation becomes harder, the right answer eventually emerges. For an example let us look more closely at Kepler's work with Mars from Chapter 1.

In that work, the German mathematician-astronomer showed that the observations were best matched if Mars followed an elliptical orbit around the Sun. Previously, it had been assumed that all the planets and the Sun went round the Earth in a circular orbit. The Catholic canon Copernicus had toyed with the idea that the Sun, not the Earth, was the centre of the Solar System, but he had retained the circular orbit. This was because circles were thought to be perfect shapes, and since the planets were in God's realm of Heaven, it follows that they should follow perfect paths. It also made the mathematics much easier because the distance of the planet from the Sun wasn't a changing quantity.

The trouble for Copernicus was that his Sun-centred model of the Solar System was worse at fitting the observations than the old Earth-centred system, so, he thought his intuition must be awry. This was the reason Copernicus was reluctant to publish his ideas: because he thought he was wrong, not because he was afraid of persecution from the church, as is often said to be the case.

When Kepler resurrected the idea about five decades later, he was so convinced that the Sun must be the centre of the Solar System that he was willing to look at everything else with a critical eye. He realized the circular orbit was simply an assumption. There was no factual basis for an orbit to be this shape, so Kepler relaxed the assumption and was competent enough to cope with the increased complexity in the mathematics. Eventually, he found that the correct orbital shape is an ellipse.

So if all the possible explanations for dark energy are running into brick walls, perhaps it is time to back up and look at the assumptions that are invisibly stitched into modern thinking. It turns out that there is a really big assumption in our standard model of cosmology and it's a fudge introduced by Einstein to simplify the mathematics. A small but persistent number of cosmologists now claim that dark energy is the product of this assumption. Take it out, do the calculation properly, and they say the need for dark energy disappears.

I spoke to Thomas Buchert from the University of Lyon, France, in 2014 about this intriguing idea. His Skype status reads 'dark energy was yesterday', and he told me in no uncertain terms, 'In five to ten years, no one will talk about dark energy.'

He is not the only one to think so. David Wiltshire from the University of Canterbury, New Zealand, works along similar lines.

The assumption is that the contents of the Universe can be averaged out into a uniform density. This allowed Einstein to solve the equations of general relativity for the whole Universe. As we have discussed, relativity equates gravity to a curvature of the spacetime continuum. If the density of the Universe can be set to a single constant value, then the curvature is the same everywhere. Einstein felt entirely justified in applying this assumption because he was working a decade before Hubble identified the galaxies as separate islands of stars. Hence, Einstein thought that stars were scattered at random throughout the cosmos.

This thinking is now so ingrained that cosmologists have raised it to the level of a scientific principle. Called the cosmological principle, it states that the Universe is isotropic, meaning that it looks roughly the same in every direction, and homogeneous, and that, by and large, its matter and energy are spread evenly throughout space. But it's not really true, as the identification of galaxies and their tendency to cluster showed.

The Coma Cluster that allowed Zwicky to make his suggestions about the need for dark matter was 25 million light years across. Another nearby grouping, known as the Virgo Cluster, was measured at more than 100 million light years in diameter. Clearly these are large departures from homogeneity – and worse was to come.

• • •

In the 1980s telescopes were not really any bigger than the ones Edwin Hubble had used, but the speed with which astronomers could take images had increased dramatically, thanks to the replacement of photographic film with digital sensors. Film is highly inefficient, failing to register most of the light that falls on it; hence, exposure times for faint celestial objects could last many hours. The circuits in the early digital cameras, which were known as charge-coupled devices, were so sensitive to light that they slashed the exposure times needed and astronomers could survey faster and deeper into the Universe than ever before. What they saw amazed them.

Clusters of galaxies were not the largest objects; instead, clusters grouped together into superclusters that stretched across space in gigantic filaments, typically many hundreds of millions of light years across. These enclose bubble-shaped voids of matter, which have just one-tenth of a cluster's density but fill more than 60 per cent of the Universe's volume.

The cosmologists called it the cosmic web. And the more they looked, the more they found. The so-called Great Wall of Galaxies was identified in 1989, spanning half a billion light years in length, a third of a billion light years in width, and 16 million light years deep. Clearly, the Universe is not as uniform as cosmologists originally thought.

To bridge this gap between the assumption and reality, cosmologists now talk about statistical homogeneity. This means that if we make the volumes we are considering the size of the superclusters and the voids, everything should average back to uniform behaviour above this scale. In the standard picture of things, this is estimated to take place on lengths of about 250 million light years.

The variation below this size will manifest itself as a perceived movement through space. Take, for example, our own Galaxy. It is gravitationally bound to more than fifty other galaxies, most of them small, in something known as the Local Group. This collection appears to be moving through space towards the Virgo Cluster of galaxies with a speed of 630 kilometres per second. Routinely, astronomers talk about this as the in-fall phenomenon.

The velocity shows up clearly in the cosmic microwave background radiation. In a uniform Universe, it should appear the same wherever we look at the sky. In reality, it is hotter in one direction by a small yet significant amount. It is called the 'dipole anisotropy' and was one of the first details to be discovered about the microwave background. It is generally explained as a Doppler effect created by Earth's own movement, which is made up of a number of components: our orbit around the Sun, the Sun's orbit around the centre of the Galaxy, the Galaxy's movement through the Local Group, and the Local Group's movement around a centre of gravity somewhere in the nearby Virgo supercluster. Take all this into account, say most cosmologists, and we are 'falling' towards the Virgo cluster at approximately 630 kilometres per second. This causes the dipole anisotropy.

But if we measure the Virgo Cluster, we detect a redshift, indicating that it is moving away from us, as we should expect in an expanding Universe. So one measurement says we are falling in; the other shows that Virgo is moving away. How do we reconcile that?

Most cosmologists will explain it in terms of the approximate way we model the expansion of the Universe. Because we know that the Universe is not homogeneous, the Einstein equations using an average density will only ever give approximately right answers. The question is: at what stage do the answers stop being approximate and start being wrong?

The spacetime continuum is not going to perform a mathematical calculation to average out its density before it decides how to expand. Each individual patch of spacetime will expand, depending upon the amount of mass within it at that particular moment. In Wiltshire's opinion, the Local Group isn't moving much at all. It is situated in a long filament of galaxies that extends through the Virgo Cluster. Since there is more matter along this filament than off to the side, the expansion of the Universe takes place at a slower rate along its length. When we look away from the filament and into the rest of the Universe, we see the expansion happening faster. This, rather than any local motion, is the cause of the dipole anisotropy: the 630 kilometres per second is the difference in the expansion rate along the filament rather than off to one side.

So, according to Wiltshire, this particular discrepancy occurs in our celestial neighbourhood, a volume of about 100 million light years across, yet it colours our view of the whole of the rest of the Universe. The deeper we look into space, the worse the problems become, because astronomers continue to discover larger and larger structures. The largest collections of galaxies in the Universe are billions of light years in length. The so-called SLOAN Great Wall extends 1.38 billion light years. In 2013, the Hercules–Corona Borealis Great Wall was identified at 10 billion light years long, 7.2 billion light years wide, and 0.9 billion light years thick.

There is also no doubt that astronomers are seeing more and more anomalous motions. In 2008, Alexander Kashlinsky of NASA's Goddard Spaceflight Center in Greenbelt, Maryland, and colleagues hit the headlines with an analysis of the Doppler shift of 700 galaxy clusters that showed they were all engaged in an apparently unexplained march through space in approximately the same direction as the dipole anisotropy.*

Kashlinsky called this motion the 'dark flow' and suggested a highly exotic reason for it. He thought that it could be caused by the gravitational influence of matter created in the very first moments of the Big Bang, but driven far over the horizon of our observable Universe by the runaway expansion of the unproven inflation hypothesis. Heady stuff.

But could this be differential expansion rearing its ugly head again? It is impossible to distinguish a Doppler effect caused by motion from a cosmological redshift caused by the expansion of the Universe. So the dark flow could be showing us something similar to the dipole anisotropy across a deeper, larger swathe of space. Instead of galaxies being pulled *through* space by a gigantic gravitational force, the expansion of the Universe is being impeded because of the amount of matter situated along that line of sight.

This may seem like a subtle distinction but the implications mount up.

By modelling the Universe as a place of overall uniform density, the anomalous movements of galaxies need to be explained. In the 1980s, astronomers used to talk about the Great Attractor. This was a supposed concentration of galaxies situated some 200 million light years away. The trouble

* A. Kashlinsky, F. Atrio-Barandela, D. Kocevski, H. Ebling, 'A Measurement of large-scale peculiar velocities of clusters of galaxies: results and cosmological implications' [arXiv:astro- ph/0809.3734]

was that it lay behind the bulk of our Galaxy, making it almost impossible to study.

In 2005, astronomers finally used X-ray telescopes to observe the Great Attractor and found that it wasn't so great after all. It contained approximately half the mass needed to cause the observed movement. But there appeared to be a large supercluster of galaxies even further away that might fit the bill. Then Kashlinsky raised the spectre of the dark flow and claimed that we would never be able to see the mass responsible because it lay beyond the boundary of our observable Universe, making it impossible to test this hypothesis. Only in billions of years time would the light from this supposed supercluster of galaxies reach the Earth. Of course, supporters of this hypothesis would argue that they cannot be held responsible for the Universe failing to collaborate in our view of ideal scientific practice.

Then there is the enigmatic 'cold spot' in the cosmic microwave background radiation that I mentioned at the opening of the book. It was seen in the first year's worth of data collected by NASA's WMAP probe as a patch of sky in the direction of the southern constellation Eridanus that looked cooler than the rest. Some thought it was a statistical fluke that would even out as more data was collected, but when Planck saw it as well, there could be no doubting its reality any longer. It is a real feature of the Universe and therefore needed explaining. It is 70 millionths of a degree colder than the average temperature of the Universe and that strains the very limits of what might be expected from inflation; the average fluctuation across the rest of the sky is just 18 millionths of a degree.

The cosmologists who believe the Universe is fundamentally

inhomogeneous would argue that such a cold spot is possible if the Universe is less dense overall along that line of sight. This would lead to a faster than average expansion in that direction, meaning that the microwaves would be stretched more and this redshift makes them correspond to lower temperatures.

In this view, the anomalous motions of galaxies across the Universe and the cold spot on the microwave background have a perfectly simple origin: they are mapping out lines of sight in which the Universe is expanding at different rates owing to a different density of matter along them. In other words, the inhomogeneity means that we have lost the isotropy as well, and the cosmological principle is false. To regain an unbiased view, we must correct for the special sight lines that we have. This is more easily said than done.

To drop the uniform density of matter assumption is to open up an infinite number of possible matter distributions, and that complicates the mathematics hugely. Thomas Buchert has no fear of doing hard sums and finds it difficult to believe the stubborn adherence to the standard model of cosmology when there is such overwhelming evidence against homogeneity.

He told me, 'In all other disciplines of physics, if you came up with a problem like dark energy, you would improve the model. This has not been done in cosmology.'

So he's done it. Buchert's model uses the selfsame equations of Einstein's that everyone agrees on, but assumes that space is fundamentally divided into two types of region: clusters and voids. Rather than calculating one average curvature for everything, he computes the curvature in each region, adds them together, and only then averages.

He finds that the Universe no longer expands equally in all directions. A nearby void on one side of the sky makes that region expand faster, whereas the higher gravitational pull in a rich cluster of galaxies will make that area expand more slowly. But the real benefit of the model is when he considers the way gravity has sculpted the Universe during the last 13.8 billion years.

As the microwave background tells us, the Universe was once extremely uniform. As time marches on, so gravity pulls matter together to form the cosmic web. In this inexorable process, it leaves large voids behind. While the expansion of the Universe has been halted inside clusters of galaxies because of the density of matter there, the voids have been unfurling ever faster. Buchert and others have published results in which they claim to show that this would give rise to an accelerating effect, rather like that ascribed to dark energy – but without a hint of the stuff in sight. He says that it is a natural behaviour, based entirely on a more precise treatment of Einstein's work.

If that is the case, then there are big consequences. David Wiltshire has already got to grips with some of these and has used a number of papers to present his concept of the Universe as a timescape.*

This idea follows quite logically from one of Einstein's masterstrokes, which was to show that space and time are conjoined entities, now referred to as the spacetime continuum. If the spacetime continuum expands at different rates in different places, it is inevitable that time will pass at different rates in different parts of the Universe too. Measure the age of the Universe from within a dense cluster and you will get one

* arXiv:0912.4563 [gr-qc]

value; measure it from within a void and you will get another. In other words, there is no such thing as a single age of the Universe; it depends on how fast your 'clock' has ticked since the Big Bang, and that depends upon the density of matter around you. It could vary by around 25 per cent, depending upon whether you are in a cluster or a void. Wiltshire's work has suggested ages of 13.7 billion or 18.6 billion years old could be measured. Either is wrong; both are right.

So, all our fundamental measurements of the Universe are contingent upon our place in the cosmos. Just as observers in chapter 5 watching lightning strike a train had to take into account their relative velocity before they could agree on whether they had seen the same event, so we must take into account our position and the density of matter around us.

Most cosmologists think that this is overstating the problem and that while there are inhomogeneities in density, they are effectively insignificant. They would prefer to continue with the standard model and look for new physics to explain the mysterious dark energy rather than complicate the calculations. There is good reason for taking this point of view, because if not even the age of the Universe is a given, it raises the question: what is time?

Similarly, we might ask the same about space, because if string theory is correct then it means there are more than the three simple dimensions we know from everyday experience. In Chapter 7, we learnt that string theory may be able to explain all fundamental forces if particles are really knots of energy.

The puzzle is that, to get the mathematics to work, theoreticians had to include extra dimensions. We only experience up–down, left–right, in–out, so these extra dimensions had to have either collapsed to almost insignificant proportions, or be blown up to enormous size, in which case our Universe was a small island in a much larger 'multiverse'.

By 1995, mathematicians had discovered that string theory comes in five distinct varieties, each one a description for a possible universe. Again, it made ideas about a multiverse look plausible. Further mathematical investigation only strengthened the idea by showing that each string theory was a special case of an overarching 11-dimensional theory, which is now referred to as M-theory. Some suggest that the M stands for 'mother', though no one seems to know for sure.

This seemed to corroborate the idea that there could be many different universes spread across a 'bulk' of higher-dimensional spacetime, and cosmologists soon began to talk enthusiastically about the multiverse. The attraction is obvious because, philosophically, it could solve the really big problem of why our Universe is like it is.

As we uncover the laws of physics one by one, most physicists do not stop to ask the question why our Universe is the way it is. Why does gravity pull with the strength it does? Why are there four (known) forces of nature; why not just three, or five? Why are electrons and protons so different in mass? Those who have played around with the laws of physics have discovered how finely tuned our Universe is in certain circumstances.

For example, if you make the strong nuclear force a little stronger, stars might never form because you can never liberate the energy. The matter would just collapse and become a black hole. If the pull of gravity were to increase, then stars would burn out more quickly – perhaps too quickly for life to form on any surrounding planets.

This raises the question: how did our Universe become so perfectly right for us to exist? The conundrum goes away if there is a multiverse, because then every possibility is tried out somewhere in the bulk. Universes come and go, and all have different physics because string theory is so flexible that many different solutions are possible. Naturally, we exist in the universe that happens to be right for us. Some cosmologists even argue that if you do not believe in the multiverse then the only recourse is to believe in a god who chooses the correct cosmos for us.

Yet there are no guarantees that string theory is right, and worryingly there seems no way to test the concept. The very mathematics of the theory makes it impossible to see these other realms. So, as with inflation, scientists have developed an unfalsifiable hypothesis – one that cannot be tested – leaving us stranded in the desert of mathematics with no apparent link to reality. Perhaps worse is that these ideas are routinely taught as if they are correct.

On 22 November 2012, a paper appeared on the arXiv server from Mikhail Shifman, professor of physics at University of Minnesota, Minneapolis.* Instead of presenting results, it was a reflection on a conference he had attended the month before. Clearly sceptical of string theory and supersymmetry, he noted that most of the young researchers in the field were spending their time working to shore up the old ideas, and he entitled the last section of his paper, 'A lost generation?'

In it, he first estimates that there are probably 2,500–3,000 particle theoreticians at work, and then he writes:

* arXiv:1211.0004v3 [physics.pop-ph]

The majority of them are young theorists in their thirties or early forties. During their careers many of them never worked on any issues beyond supersymmetry-based phenomenology or string theory. Given the crises (or, at least, huge question marks) we currently face in these two areas, there seems to be a serious problem in the community. Usually such times of uncertainty as to the direction of future research offer wide opportunities to young people in the prime of their careers. It appears that in order to take advantage of these opportunities a certain amount of reorientation and re-education is needed. Will this happen?

If there is one place that our ideas always seem to run aground, it is when we get to the very smallest dimensions of the Universe. Any theory of everything must also tell about the nature of spacetime itself. This is as close as science ever gets to asking that ultimate question: what is reality?

It is the final frontier, and is where we are going next.

10

Solving the Singularity

For all of general relativity's great successes, it leaves open the question of what spacetime is. This question of what is space and time has occupied some of the greatest minds in history, from the ancient philosophers to the scientists of the Enlightenment and beyond. Yet after thousands of years, there is still no consensus. Perhaps strangest of all is the fact that our failure to understand time has proved to be almost superfluous to making scientific progress. Newton's laws of motion, Einstein's relativity, and quantum theory do not offer us any clue about the true character of time, nor do they seem to require us to know exactly what it is.

Two possibilities exist. Each offers insights, yet they appear mutually incompatible.

The first idea is that space and time are real things, in the same way that mass or electrical charge is real. It was the viewpoint of Isaac Newton that time flows from one second to another like a river, with events ordered in sequence. It would continue to flow even if the Universe were completely empty. Likewise, space is real too. Objects are placed in it like ornaments on a shelf. Newton thought of space and time as an absolute scaffolding for the Universe and used the idea to derive his laws of motion and gravity. But the fact that space and time were virtually imperceptible put him in a bind.

He of all people knew that, without measurement, an idea

simply wasn't scientific. Clocks, for example, do not measure time but the rate of change of a mechanical or physical system. This can be the tick-tock of an escapement, the oscillation of a quartz crystal, or the ejection of a particle from a radioactive atom. Similarly, he reasoned that planets must be moving relative to absolute space, but since we cannot perceive this, we are forced to measure a planet's motion relative to something else: for example, around the Sun, or with reference to another planet.

He began to wonder if there was any way to infer the existence of absolute space and time through measurement. He imagined the rotating Earth. We know it is rotating because when we look outwards, the stars wheel across the night sky and the Sun moves from horizon to horizon. Newton wondered whether, if you took away the stars, the Sun and everything else in the Universe apart from the Earth, would we still know that we are rotating?

His answer was yes, because the rotation of the Earth causes the planet's equator to bulge. This is a manifestation of the centrifugal force, the apparent force that springs up in rotating objects that we met in Chapter 1 when talking about why the planets formed in a disc. Newton believed the appearance of a centrifugal force was the hallmark of absolute rotation. So in an otherwise empty universe, a planet would still bulge at the equator because it was rotating, even if there were no other celestial objects to measure this rotation against.

But he ran into the problem of falsifiability; in other words how do we test this idea? There is no way to do this experiment because we cannot empty out the Universe. So absolute space and time must remain a philosophical point of view on which Newton perched his science.

The second point of view is that space and time constitute an illusion. They only have meaning when we compare objects and events to one another. This was the viewpoint of Newton's arch-rival Gottfried Leibniz. He thought of time and space as purely relational concepts, similar to the way that two people can be brothers though there is no meaning to the concept of brotherhood without the people existing beforehand. Einstein also subscribed to this point of view, going so far as to say 'time is nothing but a stubbornly persistent illusion'.

Einstein was greatly influenced by an Austrian physicist called Ernst Mach, who lived across the turn of the twentieth century. Mach believed that all motion was relative and therefore that some physical law must exist between us and the fixed stars that creates centrifugal force. To Mach, if you emptied the Universe, there would be no inertia and no centrifugal force. In other words there would be nothing by which to measure your state of motion. But again, there is no way to do this experiment, so Einstein had to use it merely as a philosophical underpinning.

It drove him away from absolute space and time, and to the crowning concept of general relativity: the spacetime continuum. As we discussed in Chapter 5, this joins space and time together into a single conceptual landscape that helps us make sense of the Universe. To Einstein, spacetime was nothing but a means to an end. It didn't really exist; it was a set of mathematical relationships between the various celestial objects. But cracks appeared early in this interpretation.

Dutch mathematician Willem de Sitter showed that Einstein's equations could describe a universe that was devoid of matter. To Einstein, this was unthinkable because he derived the equations under the assumption that space and time were relational quantities only; there could be no space without two objects to measure distances between. Yet de Sitter showed there could be.

More contention came with Lemaître's realization that the Universe could expand. Even today, the expansion of the Universe is not at all straightforward to interpret, and although cosmologists argue about the meaning of it, most agree that there is one phrase in particular that should be banned: 'expanding space'.

They are happy to say that the Universe is expanding but are equally emphatic that space itself does not expand. It is a subtle distinction, but to grasp the argument leads to a great conceptual change in how to think about the Universe, almost certainly the way Einstein himself thought about things.

The idea they are trying to avoid is the implication that space is a thing that can exert a force on celestial objects and push them apart. There is a well-used analogy that likens the expanding Universe to a currant bun. The currants are the galaxies and they are driven apart during baking by the mixture rising. At first sight this might seem perfect, but it is not correct because the rising dough pushes the currants apart whereas space does not push the galaxies apart.

Another example is the frequently asked question of whether the expansion of space is taking place within the Solar System. Are the orbits of the planets getting larger with time? The answer is no, but saying that the gravity of the Solar System resists the expansion of space is only half right. It makes it sound like a tug-of-war, with the expansion of the Universe trying to pull celestial objects apart while gravity is trying to pull them together. This is not the situation at all; neither is it correct that space is expanding beneath the celestial objects, like a frictionless expanding ice rink sliding underneath a stationary person's feet.

Any thought of gravity or expansion acting like a force is classical Newtonian thinking. In general relativity, Einstein does away with these notions. Expansion is a possible behaviour of spacetime and so is gravity. Which one is manifested all depends upon the amount of matter and energy deposited in it.

Nevertheless, how can something that is nothing expand? Does more nothingness come into existence to broaden the gap? Or is spacetime a something after all? Whatever we do to try to solve this problem, the underlying problem remains: we simply do not know the nature of spacetime.

Our only hope is to scratch around for observational clues. As Newton pointed out in the seventeenth century, everything from the largest galaxy to the smallest scintilla of light is embedded in space and time. If spacetime is a thing, it must surely interact in some way with whatever is passing through it and that must imprint a signature. In principle, we could measure and interpret this to reveal the true nature of spacetime and gravity. In doing so, it would reveal the supposed quantum theory of gravity to us.

During the night of 20 June 2005, on La Palma in Spain's Canary Islands, a series of giant telescopes known as MAGIC (Major Atmospheric Gamma-ray Imaging Cherenkov) telescopes detected a burst of gamma radiation from the giant black hole that lurks at the heart of Markarian 501, a galaxy some 500 million light years away. In itself, this was not usual; any time something falls into a black hole, a flare of radiation will be given off. But detailed analysis revealed something odd. The lower-energy radiation within the burst seemed to have arrived up to four minutes before the higher-energy radiation.

This raised eyebrows because in relativity's smooth spacetime, all light travels at the same speed regardless of its energy. In quantum gravity, however, spacetime would not be smooth. It would be subject to Heisenberg's uncertainty principle and this would render spacetime a turbulent quantum foam, with no clearly defined surface. In effect, Einstein's smooth landscape would become like a choppy seascape through which particles and radiation must fight their way. Lower-energy light has longer wavelengths and would be akin to an ocean liner, gliding through this foamy quantum sea largely undisturbed. Light of higher energy and shorter wavelengths, on the other hand, would be more like a dinghy battling through the waves.

Such an effect had been suggested in 1998 by John Ellis of CERN, near Geneva, Switzerland, and Giovanni Amelino-Camelia of Sapienza University of Rome, Italy.* By scanning light from a far distant galaxy, even a subtle effect would build into a detectable time lag. On the face of it this was exactly what MAGIC had seen, but when a similar gamma-ray telescope, called HESS (High Energy Stereoscopic System) and located in the outback of Namibia, caught sight of another giant flare in July 2006, things went wrong. The galaxy in question, PKS 2155-304, is four times further away from Earth than Markarian 501, and so the delay effect should

* G. Amelino-Camelia , John Ellis, N. E. Mavromatos, D. V. Nanopoulos, and Subir Sarkar, 'Tests of quantum gravity from observations of γ-ray bursts', *Nature* 393, 763 (1998).

have been even bigger. But the astronomers saw not even a hint of a time delay.

This strongly suggests that the original effect was something intrinsic to the source of the gamma rays in Markarian 501 – possibly the acceleration of particles along magnetic fields near the centre of the galaxy, which would naturally result in the emission of lower-energy gamma rays first.

Then, on 27 April 2013, the Universe dealt a new card. A short, intense flash of radiation from the explosive death of a hypergiant star arrived at Earth's orbit. Known as a gamma-ray burst (GRB), it triggered the detectors on NASA's orbiting Fermi space telescope. GRBs are so bright that they can be seen across the entire Universe, meaning that their light has travelled through spacetime for billions of years. This latest one, catalogued as GRB130427A, was no exception. The explosion had taken place 3.6 billion light years away, and it was extraordinarily bright, deluging Earth with gamma rays of all wavelengths.

Automatic alerts were sent out to observatories across the world and within hours a battery of telescopes was observing the burst's aftermath. Amelino-Camelia and colleagues analysed the data and circulated a paper claiming to see a time lag of hundreds of seconds between the lower- and higher-energy gamma rays.* The important part was that they found a simple mathematical formula relating the time lag to the energy. This made a comparison with the predictions of various quantum gravity hypotheses much easier.

Each approach to quantum gravity sketches a different picture of spacetime, which can have quite different effects on light. In string theory, for instance, quantum spacetime is

* arXiv:1305.2626v2 [astro-ph.HE]

a tangle of six extra dimensions of space, in addition to the usual three of space and one of time. Photons of different energies will propagate through this arrangement in quite a different way from that predicted in loop quantum gravity, another popular theory that imagines spacetime as a form of chain mail composed of interwoven loops. But just when progress seemed possible, other researchers published results from other GRBs that showed no delay.

With no consistency, there can be nothing to implicate spacetime. The continuum cannot pick and choose which events to impress itself upon; either it affects them all or none at all. By extending this idea, we can see that it is not just light that would be affected. The same must hold for particles too, such as neutrinos, the ghostly particles that travel at almost the speed of light.

Although they hardly interact with matter, neutrinos must interact with spacetime because they carry energy. That means they should suffer an energy-dependent time lag too. The IceCube neutrino detector came online in 2011. Buried in a cubic kilometre of Antarctic ice, the $279 million detector swiftly registered a brace of neutrinos that appear too energetic to be coming from the Sun, by far the greatest source of neutrinos that we detect. In a fit of scientific whimsy, the neutrinos were dubbed Bert and Ernie, after two characters from the TV show *Sesame Street*.

More recently still, IceCube announced the discovery of a further twenty-six neutrinos whose energies possibly betray a distant source. More are showing up in the data all the time. The key test will be if they can be traced to a specific source, and shown to have a time lag across their energies. This work is ongoing.

Alongside all the particles and photons, there may be another interaction with the spacetime continuum that we experience directly every day. We think nothing of it, yet it has vexed some of the greatest minds in physics and may be the very clue we need into the spacetime continuum. It is inertia, a body's resistance to acceleration.

Newton successfully equated inertia to mass in his second law of motion, which states that the force at work can be computed by multiplying the object's mass by its acceleration. But his laws are silent on the mechanism by which inertia is generated. One thing is for sure: it is not coming from the Higgs boson.

In 2013, the Higgs exploded into the public consciousness when evidence for its discovery was presented by physicists sifting through the debris of particle collisions at the Large Hadron Collider at CERN. The Higgs boson is the messenger particle of the Higgs field, which permeates the Universe. It interacts with fundamental particles such as electrons and quarks to give them their masses. Yet despite being touted as the 'giver of mass', the Higgs field is not the origin of all the mass in the Universe. Not by a long way.

When quarks combine into protons and neutrons, the resulting mass is roughly a thousand times heavier than the combined mass of the constituent quarks. This extra mass comes not from the Higgs mechanism, but from the energy needed to keep the quarks together. Called binding energy, it turns itself into mass in accordance with $E = mc^2$. Somehow, these two effects combine and then latch on to something else to resist acceleration and thus create inertia. That something could well be spacetime but as yet there is no real hypothesis for how this interaction might take place.

As for time, there are unexplained things there that could be clues to progress. We experience time as a flow from past to present to future, and call this the arrow of time. The only law of physics that hints at this behaviour is the second law of thermodynamics. Its origins spring from the beginning of the nineteenth century. Nicolas Léonard Sadi Carnot, a French military engineer and physicist, was investigating the efficiency of steam engines. Although he was working at the tail-end of the industrial revolution, he was ahead of his time because his work lay fallow for decades until Rudolf Clausius and Lord Kelvin used it to help formalize the science of thermodynamics – the flow of heat.

Crudely, Carnot realized that heat only flows from hot to cold. This one-way street in physics leads to irreversible processes. For example, a hot cup of tea placed in a room goes cold because it naturally transfers its heat into the air. Once the tea is at the same temperature as the room, the transfer stops and it is impossible for the process to reverse.

The thermodynamic energy, which is now spread throughout the room, never finds its way back into the cup to warm it up. Here we see time's recognizable arrow. If we see a film of a plate smash, we know that time is 'running' forwards. We instantly recognize if the film is run backwards because we know that plates never spontaneously reform from fragments. That would require the flow of energy to be reversed and this simply does not happen in our Universe.

By the middle of the nineteenth century, this behaviour had been linked to a concept called entropy. It can be thought of as a measure of the disorder in a system, and it never decreases. A smashed plate is at higher entropy than a whole one. To put that plate back together would require the expenditure of energy that increases the entropy in our surroundings. It is a no-win situation. You can never decrease the overall entropy of the Universe because everything that happens is driven by the one-way flow of energy from hot to cold. This certainly makes it look as if time is flowing in a single direction, but it still doesn't answer the basic question: what is time?

Help may be at hand from a particularly intriguing correlation that was first noticed by computer scientists during the Second World War.

Time is clearly linked to the concept of entropy, the measure of the disorder in a system. But entropy is more than that. It crops up in the relatively modern field of information theory.

Information theory as a science can be pegged to a landmark paper published in 1948.* It was written by Claude E. Shannon, an American cryptographer who worked at Bell Labs, New Jersey. During the course of the Second World War, Shannon worked on the American codebreaking effort, spending time with the great British codebreaker Alan Turing and realizing that many of their ideas were complementary.

Using a purely scientific approach to codebreaking, meant that Shannon needed to quantify the messages he was working with, and that led him to develop a way of measuring the information content of a message. He wrote numerous classified documents during the war, and then during peacetime put the central arguments together in the form of his 1948 paper, 'A Mathematical Theory of Communication'.

* Shannon, Claude E. (July-October 1948). 'A Mathematical Theory of Communication', *Bell System Technical Journal* 27 (3): 379-423. doi:10.1002/j.1538-7305.1948.tb01338.x.

A key concept, which is now familiar to everyone involved with computers, is that the most fundamental unit of information is the bit. This can exist in one of two opposing states: one or zero, yes or no, left or right, an electrical pulse or no electrical pulse. In modern life we probably come across this most often when contemplating our broadband providers, who all advertise their speed in megabits per second, i.e. millions of bits per second.

Shannon called the amount of information a message can hold its 'entropy', and people soon started noticing that thermodynamic entropy and information theory entropy were closely related, perhaps even exactly the same thing. One of those people was Steven Hawking.

He was working on the search for a quantum theory of gravity, and his calculations concerned the microscopic behaviour of black holes. It was well known that matter falling into a black hole is lost forever, but a number of researchers had realized that this does not work from a thermodynamic point of view.

Matter always contains entropy, and according to the second law of thermodynamics entropy can only ever increase. If matter disappears into a black hole, what happens to that entropy? Is it lost along with the matter? If so, the entropy of the Universe would decrease and black holes would violate the second law of thermodynamics. Stephen Hawking thought this was fine. It never worried him to discard a concept if it stood in the way of progress. Others weren't so cavalier.

Jacob Bekenstein had just finished his thesis. He had been studying at Princeton under the great American physicist John Wheeler, who is often erroneously credited with coining the term 'black hole'. Wheeler had asked Bekenstein to look into the loss of entropy problem, and by 1972 Bekenstein had a possible solution.

Black holes hide their singularity within a boundary known as the event horizon. Nothing that crosses the event horizon, including light, can ever return to the outside Universe. Bekenstein realized that the size of this horizon was the key to the problem. When matter falls in, the horizon grows in response to the consumption of entropy.

For some reason, Hawking did not like this answer. At a 1972 summer school in the French ski resort of Les Houches, Hawking and two colleagues, Brandon Carter from Meudon Observatory and James Bardeen from Washington, confronted Bekenstein and argued with him. After the conference, Hawking set about disproving Bekenstein's work. It took him a couple of years to complete his numerical deliberations, and although it was not the answer he expected, it was Hawking's greatest breakthrough.

Instead of scotching his young rival's work, he confirmed it by discovering the precise mathematical form of the relationship between entropy and the black hole's event horizon. As a result Hawking made a complete U-turn, and embraced the idea that thermodynamics played a role in black holes. Then he went further.

Anything that has entropy, he reasoned, also has a temperature, and anything that has a temperature can radiate energy. He had heard from researchers in Russia that they were finding theoretical hints that black holes might not be completely black – that the action of quantum mechanics near the black hole event horizon could lead to the apparent emission of particles from the black hole. Hawking showed how this could happen in a paper to *Nature* in 1974.*

He proposed that matter could appear to escape under certain conditions from a black hole and that this would eventually lead to its complete evaporation. He provocatively entitled the paper 'Black Hole Explosions?' and the radiation is now called Hawking radiation. The work led to a new bet. As with the first, it involved Kip Thorne, but this time, instead of betting against each other, they teamed up and bet another researcher, John Preskill. The winner would receive an encyclopedia of their choice.

Hawking and Thorne believed that the emitted radiation was not associated in any way with the matter that had fallen into the black hole. They imagined it was produced just outside the black hole and carried no information about the interior. Whatever information fell into the black hole, stayed in the black hole, effectively destroyed.

Preskill thought that was nonsense. Information could not simply be destroyed because, according to quantum mechanics, it should be conserved. The puzzle became known as the black hole information paradox. The bet was made in 1997 and ran for seven years. It was finally conceded by Hawking in 2004. His further studies had convinced him that the radiation coming out of the black hole *was* related to the matter and radiation that had fallen in. He bought Preskill a copy of *Total Baseball: The Ultimate Baseball Encyclopedia*. Thorne, however, declined to contribute to the prize because he was unsure Hawking had really solved the problem.

In truth, the bet is simply a sideshow; the most important thing to come out of the work was the original equation

* 'Black hole explosions?', S. W. Hawking, *Nature* 248, 30 - 31 (01 March 1974); doi:10.1038/248030a0.

found in the 1970s for the black hole's entropy. This is now called the Bekenstein–Hawking entropy equation and is the one Hawking has asked for on his gravestone. It represents the ultimate mash-up of different physical disciplines because the equation contains Newton's constant, which clearly relates to gravity; Planck's constant, which portrays quantum mechanics at play; the speed of light, which is the talisman of Einstein's relativity; and the Boltzmann constant, which is the herald of thermodynamics.

The presence of these diverse constants hints at a quantum theory of gravity, which could then be united with all the other branches of physics. Furthermore, it strongly corroborates the feeling that the understanding of black holes will be key in unlocking that deeper theory. And it points to the fact that entropy and information are important components of whatever our final theory will be.

Some physicists are willing to take this further and postulate that information may be the more fundamental thing in the Universe rather than matter and energy. This idea has led to the holographic principle, developed by Dutch physicist Gerard 't Hooft and expanded upon by American physicist Leonard Susskind. The name derives from the fact that holograms store information on a two-dimensional surface, but the information itself encodes a three-dimensional structure. The analogy with a black hole seems perfect because Bekenstein showed that the surface area of a black hole, which is effectively two-dimensional, correlates with the amount of information in the black hole – this is exactly the behaviour of a hologram.

Carrying this idea forward, it could be that by altering our point of view to think in terms of information, we may gain the new perspective we need to see deeper. Remember how the Earth was originally thought to be stationary because the Sun moved through the sky? This explained day and night, and the seasons, but could not explain how the planets moved through the night sky. Copernicus and others had the insight that motion is relative and so perhaps the Earth moved and the Sun was stationary. With this key change in perspective, planetary motion could be understood within a framework that also made sense of night and day and the seasons. Maybe we can do the same trick with cosmology.

Changing our viewpoint in this way is akin to changing our underlying philosophy of how we approach the problem, and this is significant because a change of philosophy often lies at the heart of a fundamental advance in science because it has changed the way we look at the problem. Indeed, the scientific revolution itself was sparked by a change in philosophy brought about by a French philosopher's feverish dream.

Back in the seventeenth century, even bad news travelled slowly. It was 1633 and that summer Galileo had been hauled to the Vatican to be condemned to spend the rest of his life under house arrest. His crime had been the vigorous promotion of the idea that the Earth moved through space, orbiting the Sun.

To the Roman Catholic Church of the 1600s, such actions were seen as a threat to its authority, and so Galileo was called to account for his actions. The drama unfolded in the sweltering June heat, but news did not reach Descartes in the Netherlands until the frozen month of November. By then, it was almost too late.

He was thirty-seven and on the verge of publishing an epic

description of a new way to investigate nature. At the heart of this masterwork was a discussion about the origin of the Sun and the planets which assumed, as Galileo had, that the great orb of the Sun was the centre of things. He planned to call his thesis *The World*, the name used during the seventeenth century to mean the entire Universe. The writing alone had taken him four years but the full endeavour stretched back fourteen winters, to another bitter night, this time in Germany.

Europe at that time was divided along religious lines, with Catholics fighting Protestants in what was to become known as the Thirty Years War, and Descartes was a young man hungry for experience but adrift with no clear sense of direction. He was drafted into the Catholic army of Duke Maximilian of Bavaria, and on the night of 10–11 November 1619 was in temporary lodgings at Neuberg. Holed up in a room with a stove for warmth, he experienced a fitful night in which three dreams changed his life.

In the first, he was menaced by dark shadows and blown from his feet by the wind. In the second, he was jolted awake by a sound that roared like thunder. Then came the third dream, which had entirely a different character; in his subconscious musings he found a dictionary and a book of poetry lying on a table. He knew the dictionary would be useful to his studies but it was the poetry that he opened. His eyes fell upon the opening line *Quod vitae sectabor iter?* (What path shall I follow in life?). While contemplating this question, a kindly stranger appeared in the dream and handed him a sheet on which was written some verses beginning *Est et non* (It is and it is not). Upon waking, Descartes claimed to know that the future direction of his life was to find a way to pursue true knowledge of nature.

The nocturnal visions may not have been the only reason for his sudden sense of purpose. Earlier the previous day, he had met Isaac Beeckman, a Dutch philosopher, who had told him of the certainty with which measurement could be applied to the investigation of nature. This was a revelation to Descartes. It implied that if something could be measured, something true could be known about the object of study: perhaps a length or a position or a speed.

Beeckman was also an early subscriber to the belief that the Universe was some sort of giant clockwork mechanism and that interaction was transmitted by contact. Descartes later recalled how he had drifted to sleep on that fateful night 'full of enthusiasm, carried away completely by the thought of having discovered the foundations of a marvellous science'.

Now, fourteen years later, he was ready to present to the world a new philosophy built on those foundations. To begin with, Descartes discarded any belief that could not be known with absolute certainty. He then attempted to build up a system of thought and investigation that would allow one to establish what was real about the world and the wider Universe. All of it was based upon the principles of mechanical philosophy. The crowning glory was the reasoned argument that the planets orbited the Sun because they were being swept along by a vortex of ethereal particles. It was perfect mechanical philosophy: movement through contact.

Descartes went further, as discussed in chapter 1, and argued that the planets themselves must have originally condensed out of the ether, like clouds condense out of the atmosphere of Earth. Then the news of Galileo arrived.

The chill the French philosopher felt must have been every bit as icy as the wind that was freezing the Dutch canals at that time of year. Catastrophically, the very foundation stone to which he had anchored his planetary ideas was Galileo's theory of the moving Earth; and this thinking was now tantamount to heresy. Although Descartes was living in a Protestant country, he remained devoutly Catholic and, on the eve of sending the book to the printers, he withdrew his masterwork and hid it.

During the subsequent years, he recast much of his philosophy and mathematics in more general terms and these he did publish. In 1637, he showed how the positions of things in the real world could be placed on a graph with numbered axes – now called the Cartesian coordinate system. That same year, he published the famous phrase 'Je pense, donc je suis' ('I think, therefore I am').

But it wasn't until 1677 that his idea on the origin and movement of the planets was finally published. By then, he was safe from persecution; he had been dead for twenty-seven years, the sparks of his philosophy had lit fires across Europe, and the scientific revolution had begun.

A contemporary of Descartes was the English philosopher Francis Bacon. In 1620 he published the *Novum Organum* (The New Method), in which he described a way of gaining knowledge based upon observation. From these observations, he suggested that laws could be built to explain the way nature behaved. As more observations were made, so the laws could be generalized to explain more and more. In this way, the laws of physics could be built up step by step, expanding the umbrella of their explanation.

But the history of physics has shown that things are never quite that simple. Sometimes new observations are not enough to make progress. Instead they highlight the limits of an old way of thinking. When this happens, the old ways must be replaced, but to do that, new underlying assumptions and philosophy are necessary.

The central example of this is the development of relativity. Einstein was forced to imagine gravity in a completely different way from Newton. Instead of a thread of force connecting two objects, as Newton thought of gravity, Einstein saw a landscape of spacetime. This radical shift of view greatly impressed the Austrian-born British philosopher Karl Popper. He was also struck by the fearless nature of Einstein's proposal, specifically the way that relativity stood or fell on the prediction of starlight's deflection around the Sun. This was a phenomenon that Newton's gravity simply could not explain. No attempt to massage Newton's work was going to make that observation fit; it needed a whole new way to think about the Universe.

It led Popper to conceive the idea that science did not proceed by proof but by falsification. A theory could never be proven; it could only be useful at explaining observations until it was shown to be false. At that point it had to be superseded by a more complete theory.

He compared the theory of relativity to the supposedly scientific work of Freud and others at the time, and found a profound difference. The work of the psychoanalysts, he thought, was sadly lacking because there was no way, even in principle, that you could falsify their ideas. To Popper, that did not deserve to be called science.

Slightly later in the twentieth century, American physicist Thomas Kuhn entered the philosophical realm by publishing his idea of the paradigm shift. This is the moment when basic assumptions change and, as a result, science looks at the world in a new way. He used an optical illusion as his analogy. There is a single drawing that can be seen as either a duck or a rabbit; two faces or a vase is another common illusion of this type. The paradigm shift was seeing one and then the other, but it is impossible to see them both at the same time.

Normal science, said Kuhn, was the day-to-day collection of facts and the interpretation of them in accordance with the accepted paradigm of the day. However, like ash at the bottom of a fire, measurements that simply do not fit build up until they become impossible to ignore. In science, they force a change in the underlying assumptions, and this imposes a change on the way we see things, and a paradigm shift takes place.

This is exactly what happened with Einstein's relativity. It may also be the coming state of cosmology today because, as we have seen throughout this book, there are growing collections of observations that simply cannot be squared with today's theories.

Eddington's method of model building is now endemic in astronomy. It is used to reconstruct the physical behaviour of most celestial objects, and even the whole Universe.

The standard model of cosmology is without doubt an extraordinary edifice of human ingenuity. It has been painstakingly put together over the last century with ever-increasing amounts of data. But it is not yet watertight. As we have seen there are discrepancies and weaknesses, and there are other ways of explaining things.

The danger with models is always that instead of attempting to falsify them, the tendency is to extend them to incorporate the tricky observations. This leads to increasingly elaborate models that can explain a great many things but, just as Popper thought about psychoanalysis, the elasticity of their design is not a strength but a weakness. These days, cosmological models are often so malleable that a 'fit' can be found just by varying the input parameters a little. This is often seen as a sign that the model must be 'more-or-less' right, but the truth is that any sufficiently general model, with many free parameters as the unconstrained quantities are called, can be used to fit virtually anything, and that makes them impervious to assault. As Steinhardt called such ideas, a theory of anything (see previous chapter).

Eddington knew that modelling should be approached with caution and its conclusions should always be viewed with scepticism. It is not clear that every modern cosmologist applies this rigour.

As we have seen, the investigation of the Universe has brought us to a point in which 96 per cent of the Universe is thought to be composed of unknown dark matter and dark energy. The inclusion of these was certainly well motivated and physically valid, but so many researchers now seem unquestioningly to believe the substances to be real rather than just possible explanations. Claims are often made that dark matter has been proven to exist, but this is simply untrue and is driven by nothing but circumstantial evidence and the lack of a better explanation.

It should always be in the back of astronomers' minds that a wrong assumption somewhere along the line could have propelled them down a fruitless avenue from which it is becoming increasingly difficult to reverse.

The sense of understanding the limits of science – and scientists – is a recurrent theme in the writing of the early pioneers. Seven centuries before Isaac Newton, the Arabic scholar Ibn al-Haytham (Alhazen) wrote:

> The seeker after the truth is not one who studies the writings of the ancients and, following his natural disposition, puts his trust in them, but rather the one who suspects his faith in them and questions what he gathers from them, the one who submits to argument and demonstration, and not to the sayings of a human being whose nature is fraught with all kinds of imperfection and deficiency. Thus the duty of the man who investigates the writings of scientists, if learning the truth is his goal, is to make himself an enemy of all that he reads, and, applying his mind to the core and margins of its content, attack it from every side. He should also suspect himself as he performs his critical examination of it, so that he may avoid falling into either prejudice or leniency.

Here we read that the mindful scientist should be distrustful of everything – even him- or herself. The parallels are clear in a heading that Newton wrote in his 1664 notebook entitled *Quaestiones Quaedam Philosophicae* ('Certain Philosophical Questions'): 'Plato is my friend, Aristotle is my friend, but my best friend is truth.'

There is clear danger in trying to prove a theory because it implicitly means that the researcher believes it to be true. This violates the prerequisite that science should be objective. When Francis Bacon was writing his *Novum Organum*, he cautioned would-be scientists about what he called the idols of the mind. He listed them in the book:

- *Idola tribus* (idols of the tribe) This is our tendency to see more order than truly exists. It comes about because people try to feed new facts into their preconceived ideas.
- *Idola specus* (idols of the cave) This weakness is due to each individual's personal likes and dislikes, which cloud judgement and reason.
- *Idola fori* (idols of the marketplace) This is confusion that comes about through the use of words in science that may have a slightly different meaning than that of common language. This also leads to confusion between individual disciplines as well.
- *Idola theatri* (idols of the theatre) This idol comes about by blindly following academic dogma and therefore not asking enough real questions about the world.

This is the true gold standard for science: constant self-questioning. And nowhere should this be applied more rigorously than when scientists build models.

One can only imagine the despair that James Jeans would feel at today's astronomy and cosmology. Back in the 1920s his disdain for Eddington's model of a star came down to there being no way to look inside the Sun and measure its internal temperature. In that particular instance, it turns out that Nature, like fairy godmothers, sometimes grants wishes. The neutrino was discovered, which allowed us to probe directly into the nuclear heart of the Sun. Today, all our wishes would be granted if we could directly detect dark matter and dark energy.

There are only two possibilities. Either dark matter and

dark energy are real and these vast reservoirs of matter and energy are just waiting to be found, or we have to radically rethink fundamental physics.

Either way, we live in an unknown universe.

Further Reading

Barbour, *The End of Time* (W&N)

Chown, *Afterglow of Creation* (Faber and Faber)

Chown, *The Magic Furnace* (Vintage)

Chown, *Quantum Theory Cannot Hurt You* (Faber and Faber)

Clark, *Is There Life on Mars?* (Quercus)

Clark, *The Day Without Yesterday* (Polygon)

Clark, *The Sensorium of God* (Polygon)

Clark, *The Sky's Dark Labyrinth* (Polygon)

Clark, *The Sun Kings* (Princeton)

Close, *Antimatter* (Oxford)

Close, *Neutrino* (Oxford)

Collins and Pinch, *The Golem* (Cambridge)

Eddington, *Space, Time and Gravitation* (Martino)

Eddington, *The Internal Constitution of the Stars* (Cambridge)

Frenkel, *Love and Math* (Basic)

Gleick, *Isaac Newton* (Harper)

Isaacson, *Einstein* (Pocket Books)

Khun, *The Structure of Scientific Revolutions* (Chicago)

Kumar, *Quantum* (Icon)

Mahon, *The Man Who Changed Everything* (John Wiley)

Popper, *The Logic of Scientific Discovery* (Routledge)

Singh, *Big Bang* (Harper)

Smolin, *Time Reborn* (Penguin)

Smolin, *The Trouble with Physics* (Penguin)

Sobel, *Galileo's Daughter* (Fourth Estate)
Sobel, *The Planets* (Fourth Estate)
Tegmark, *Our Mathematical Universe* (Penguin)
Westfall, *Never At Rest* (Cambridge)
Woit, *Not Even Wrong* (Vintage)

Image Credits

Introduction: The Day We Saw the Universe
Planck captures the Universe's first light
(credit: ESA and the Planck Collaboration)

Chapter 1: The Architect of the Universe
Saturn and its rings, captured by Cassini
(credit: NASA/JPL/Space Science Institute)

Chapter 2: Selene's Secrets
Earthrise above the Moon, as seen by Apollo 17
(credit: Image Analysis Laboratory/NASA Johnson Space Center)

Chapter 3: Gravity's Crucible
The Sun erupts, a coronal mass ejection, 31 August 2011
(credit: NASA/SDO/AIA/GSFC)

Chapter 4: The Stellar Bestiary
10 million stars are packed in a sphere just 160 light years across at the core of Omega Centauri
(credit: NASA, ESA, and the Hubble SM4 ERO Team)

Chapter 5: Holes in the Universe
A billion-solar-mass black hole lurks at the centre of Centaurus A
(credit: X-ray: NASA/CXC/CfA/R.Kraft et al.; Submillimeter: MPIfR/ESO/APEX/A.Weiss et al.; Optical: ESO/WFI)

Chapter 6: The Luxuriant Garden

NGC 1300, a barred spiral galaxy

(credit: NASA, ESA, and The Hubble Heritage Team (STScI/AURA), Acknowledgment: P. Knezek (WIYN))

Chapter 7: Chiaroscuro

Asteroid Eros

(credit: NASA/JPL-Caltech)

Chapter 8: The Day Without Yesterday

Planck's first release of the Cosmic Microwave Background radiation, obscured by dust

(credit: ESA/LFI & HFI Consortia)

Chapter 9: Timescapes and Multiverses

A computer simulation of the Universe at the largest scales: a web of dark matter filaments flecked with bright galaxies?

(credit: MacFarland, Colberg, White (Munchen), Jenkins, Pearce, Frenk (Durham), Evrard (Michigan), Couchman (London, CA) Thomas (Sussex), Efstathiou (Cambridge), Peacock (Edinburgh) /National Science Foundation /NASA.)

Chapter 10: Solving the Singularity

Galaxy cluster CL 0024+17 warps space around it, distorting the light of its galaxies

(credit: NASA, ESA, M.J. Jee and H. Ford (Johns Hopkins University)

Index